Patricia G. Patrick • Sue Dale Tunnicliffe

Zoo Talk

 Springer

Dr. Patricia G. Patrick
Texas Tech University
Lubbock, TX
USA

Dr. Sue Dale Tunnicliffe
Institute of Education
University of London
England
UK

ISBN 978-94-007-4862-0   ISBN 978-94-007-4863-7 (eBook)
DOI 10.1007/978-94-007-4863-7
Springer Dordrecht Heidelberg New York London

Library of Congress Control Number: 2012947929

Printed on acid-free paper

Springer is part of Springer Science+Business Media (www.springer.com)

Zoo Talk

Contents

Chapter 1
Introduction

Children have an inherent desire to run after butterflies, love beautiful birds and wild places, and want to make friends with elephants and tigers. (Pandey, 2003, r2)

Why does the sun set in the west? And why does my heart keep beating in my chest? . . . I got a PBS mind in an MTV world. (Buffett & Mayer, 1999)

The term biophilia was coined by Fromm in 1964. E. O. Wilson (1984) uses the term biophilia to refer to a human's innate need to be with the living world and a basic need to experience nature firsthand. Conversely, Louv (2006) believes that humans are becoming more removed from nature and suffer from nature-deficit disorder (Louv, 2006). Moreover, Kahn (2011) states that personal experiences of nature have become more vicarious and virtual through two-dimensional images and three-dimensional representations of organisms. The problem with the images and representations is that they often are reduced in size and only imitations of reality. For example, robotic dogs are being used as pets, and children take care of virtual internet pets. The good news is that there are indications that children are in touch with plants and animals in their everyday lives (Byrne, Tunnicliffe, Patrick, & Grace, 2010; Patrick & Tunnicliffe, 2011) and possess an innate desire to spend time in the natural world. Zoos offer people opportunities to interact with organisms and develop personal, authentic experiences as opposed to virtual representations and interfaces. Zoos play an important role in linking humans with wildlife.

An essential component of the zoo visit is the verbal interactions that occur among the visitors (Tunnicliffe, 1996a, 1996b). Therefore, identifying the language used during a zoo visit and how and where the linguistic interactions occur are an important part of understanding the knowledge gleaned by the visitor. Language is a complex and evolving process, which involves knowledge of vocabulary and a grasp of how words, when put together, represent things, actions, and feelings.

Language acquisition is a key stage in child development, and a child's first words are considered important by parents and caregivers. Some of the first words spoken by a child include the names of animals that the child encounters everyday (Tunnicliffe, Boulter, & Reiss, 2011). These findings are important to understanding the foundations of biological science and how zoo visits influence

P.G. Patrick and S.D. Tunnicliffe, *Zoo Talk*, DOI 10.1007/978-94-007-4863-7_1,
© Springer Science+Business Media Dordrecht 2013

learning, especially for young children. Observations, conversations, and personal experiences are the foundations of science learning. As dialogues occur between people, the people involved are coordinating and communicating their perspectives on the world. The dialogue creates an intra-perspective and inter-perspective space for understanding. Conversations develop internally (intra-perspective) and become a part of inter-perspective space, which makes conversations bidirectional. Thoughts are co-constructed within dialogue. Therefore, the preparation for and visit to an animal collection are special and essential, in terms of biosocial learning. Biosocial learning takes place during scientific events and experiences that are organized by adults for children. During the visits, adults act as facilitating dialoguers talking to the young learners who, even if they do not yet talk out loud, listen and cognitively process the language. This early stage of biosocial learning about biodiversity is experiential, but the facilitator designs the visit and provides the labels for the phenomenon observed.

There are two important voices that interact during a zoo visit, the Visitor Voice and the Zoo Voice. The Visitor Voice is guided by the interactions between children, adults, and the physical manifestations of the zoo. When young children visit zoos, they label the organisms they see and relate the in situ observations to prior knowledge and ex situ experiences. The adults accompanying the children name the organisms and provide their own interpretations. During these interactions between children and adults, children develop skills in observing the features and behaviors of animals by relating them to species with which they are already familiar. These dialogues are the voices of the visitors. However, there is another voice contributing to learning, the voice of the zoo. The Zoo Voice is guided by the exhibits, facilities, mission, staff, educators, and organisms housed at the zoo. The role of the Zoo Voice is to allow visitors to construct new concepts based on the visitors' prior knowledge. Moreover, the Zoo Voice is an advocate for organisms, biodiversity, habitat preservation, ecosystems, and biological conservation issues.

The Visitor Voice and Zoo Voice are equally important aspects of a zoo visit and must be understood separately to develop an understanding of the ways in which they influence each other. Zoo Talk discusses the two voices. Toward the end of the twentieth century and into the twenty-first century, the emphasis of zoo and aquaria research has changed from evaluation studies, such as tracking visitors during a visit and listening to spontaneous dialogues, to acknowledging that the Visitor Voice and the Zoo Voice simultaneously may be partners and rivals. The Zoo Voice is heard officially through its mission statement (Patrick, Matthews, Ayers, & Tunnicliffe, 2007). The mission statement is presented by the zoo and interpreted by the visitors through the exhibit design, amenities, facilities, and programs (Ross & Gillespie, 2009). However, visitors' ideas concerning zoos do not align with zoos' self-interpretations (Patrick, 2011).

Each visit begins with an entry narrative (Doering & Pekarik, 1996) that is based on the visitor's culture and prior knowledge. The entry narrative shapes the tone of the visit and determines what the visitor will see. The Visitor Voice during the entry narrative affects the interpretation of the Zoo Voice. In addition to the entry narrative, each visitor's agenda determines the interactions that take place within the group (Anderson, Piscitelli, & Everett, 2008). The visitors' identities, which they bring to the zoo, can evolve during the zoo visit and transform based on the groups' agenda. Therefore, the Visitor Voice is an important aspect for zoos to consider when zoos determine how they will translate the Zoo Voice.

Zoo Talk begins with a look at the Zoo Voice by briefly describing the history of zoos and providing a rationale for the existence of zoos. The Visitor Voice is more closely scrutinized as a result of identifying pertinent literature and recent research. Moreover, the book identifies the Visitor Voice by defining the public's knowledge of zoos, why people visit zoos, and their interactions within exhibits. The final chapters provide suggestions for field trip design based on the inferences made by the Visitor Voice and the Zoo Voice. The implications for the future of zoos are that zoos must listen to the Visitor Voice as well as assessing the impact of their own Zoo Voice.

References

Anderson, D., Piscitelli, B., & Everett, M. (2008). Competing agendas: Young children's museum field trips. *Curator: The Museum Journal, 51*(3), 253–273.

Buffett, J., & Mayer, J. (1999). I don't know and I don't care. *On Beach house on the moon* [CD]. New York: Island Records.

Byrne, J., Tunnicliffe, S. D., Patrick, P., & Grace, M. (2010, July). *Children's understanding of animals in their everyday life in England and the USA*. Poster presented at the ERIDOB Conference, Braga, Portugal.

Doering, Z. D., & Pekarik, A. J. (1996). Questioning the entrance narrative. *Journal of Museum Education, 21*(3), 20–25.

Fromm, E. (1964). *The heart of man*. New York, NY: Harper & Row.

Khan, P. (2011). *Technological nature: Adaptation and the future of human life*. Cambridge, MA: MIT Press.

Louv, R. (2006). *Last child in the woods, saving our children from nature-deficit disorder*. Chapel Hill, NC: Alonquin Books.

Pandey, P. (2003). Child participation for conservation of species and ecosystems. *Conservation Ecology, 7*(1), r2. http://www.consecol.org/vol7/iss1/resp2

Patrick, P. (2011, April). *Zoo acuity model: Middle level students' knowledge of zoos*. Paper presented at the National Association for Research in Science Teaching, Orlando, FL.

Patrick, P., Matthews, C. E., Ayers, D. F., & Tunnicliffe, S. D. (2007). Conservation and education: Prominent themes in zoo mission statements. *Journal of Environmental Education, 38*(3), 53–59.

Patrick, P., & Tunnicliffe, S. (2011). What plants and animals do early childhood and primary students name? Where do they see them? *Journal of Science Education Technology*. Available at: http://www.springerlink.com/content/e27121057mqr8542/

Ross, S. R., & Gillespie, K. (2009). Influences on visitor behavior at a modern immersive zoo exhibit. *Zoo Biology, 28*(5), 462–472.

Tunnicliffe, S. (1996a). Conversations with primary school parties visiting animal specimens in a museum and zoo. *Journal of Biological Education, 30*(2), 130–141.

Tunnicliffe, S. (1996b). A comparison of conversations of primary school groups at animated, preserved and live specimens. *Journal of Biological Education, 30*(30), 195–206.

Tunnicliffe, S., Boulter, C., & Reiss, M. (2011). Getting children to talk about what they know of the natural world. *Primary Science, 119*, 24–26.

Wilson, E. O. (1984). *Biophilia*. Cambridge, MA: Harvard University Press.

Chapter 2
A History of Animal Collections

Zoo managers today describe the recent renaissance of American zoos in terms much like those invoked a century ago, when zoo promoters said they were differentiating themselves from menageries by moving toward permanence, education, and science ... Adopting a new name—turning zoos into conservation parks—is a strategy to distance today's displays from the past. By emphasizing discontinuities in philosophy and exhibitry today's zoo promoters dismiss and discredit historical practices. (Hanson, 2002, pp. 163–164)

This chapter provides a brief history of the available literature on zoo education, includes a condensed review of the development of zoological institutions from menageries to conservation centers, and defines the change of collections of exotic animals from curiosities to biological conservation and education centers. Zoos moved from indicators of power to scientific establishments for taxonomy, to modern centers of conservation biology. Initially exclusive places for a chosen few to visit, zoos became accessible to the paying public. However, in a few exceptions, zoo visitation was free. Through the historical literature, we build a foundation for this book by describing the educational evolution of zoos. If zoos had not evolved from places of curiosity, they would not be centers for biological conservation and education. Additionally, the body of this book is framed by defining the roles of zoos, describing the educational experiences at zoos, including school visits and exhibitry, and explaining the importance of zoos in biological conservation education and understanding fundamental biology.

The Beginning of Menageries and Zoos

Throughout the history of animal collections, people have placed great importance on the collection of both fauna and flora (Croke, 1997; Gray, 1991; Hancocks, 2001). Historically, the purpose of what we today refer to as the "zoo" has changed to adapt to shifting attitudes toward animal confinement, to advances in technology, and to fluctuating changes in societal demographics (Turley, 2001). During the time of early humans, a relationship developed between humans and the first wolf ancestor

P.G. Patrick and S.D. Tunnicliffe, *Zoo Talk*, DOI 10.1007/978-94-007-4863-7_2,
© Springer Science+Business Media Dordrecht 2013

of the dog. The wolves scavenged for food scraps, and humans began an association with animals, which over time led to the first domesticated animal, triggering a fundamental shift in the human–wild animal relationship (Croke; Hancocks). This shift in the human–animal association eventually led to the desire to keep animals as a sign of power or human enjoyment. Today, we consider zoos to be a place for family recreation, scientific exploration, and education. The public's view of zoos is an important aspect in motivating the public to visit the zoos. Worldwide, 11% of the population visits zoos and aquariums, approximately 700 million visitors each year (Dick, 2010). The American Zoo and Aquarium Association (AZA) accredits 216 facilities that attract about 140 million visitors (Ballantyne, Packer, Hughes, & Dierking, 2007; Falk, Reinhard, Vernon, Bromenkent, Heimlich, & Deans, 2007; Wagoner & Jensen, 2010).

While some debate what could be called the "world's first zoo," the predecessor to which was the menagerie, the act of animal keeping dates back to ancient times as early as 3,500 BC in Hierakonpolis, Egypt (Dembeck, 1965; Hoage & Deiss, 1996; Rose, 2010). Previous publications date the earliest instances of "true zoo management" to the fifth dynasty of Old-Kingdom Egypt in Saqqara around 2,900–2,200 BC. The likely reason for keeping these animals in captivity was for leisure, nobility, and hunting and fighting purposes (Croke, 1997). Inherent wealth allowed employing a large staff to carry out the semidomestication and care of these animals (Bostock, 1993; Fisher, 1966). Eventually, as trade between nations increased, animals became a political pawn. Society saw the gift of an animal as powerful for both the gift giver and the receiver (Baratay & Hardouin-Fugier, 2002) and considered ownership of such animals as proof of wealth and stature.

Around 1,100 BC, the entitled Chinese emperor Wen Wang founded the first zoological garden specifically used for scientific and educational purposes. Named the "Ling-Yu," translated as "Garden of Intelligence," this 1,500-acre park displayed animals from various regions of the Chinese empire instead of exotics from foreign lands. While little record survives of the actual collection, beyond it being "extensive," it did exist around 800 years prior to the building of the Great Wall (Blair, 1929; Bostock, 1993; Fisher, 1966). It was not the only extensive imperial menagerie in Asia, as in the thirteenth century, European explorer Marco Polo described the collection of the Mongol ruler Kublai Khan.

> There was a huge marble stone and palace adjacent to a wall that surrounded many miles of parkland where game animals of all sorts were kept for hunting. Along with game animals, Kublai maintained an animal collection that included leopards, lions (probably tigers, since they were described as having black stripes), lynxes, and even elephants, as well as a variety of eagles and falcons used for hunting. (Hoage, Roskell, & Mansour, 1996, p. 12)

Both Greeks and Romans possessed vast collections of wild animals, with the Greeks focusing on songbirds (passerines). Priests and followers associated many species of birds with certain gods and therefore held the birds as sacred. The Greek public traveled for miles to view birds from India. This may be the earliest recorded instance of paying a gate fee to enter a menagerie (Hoage et al., 1996). Conversely, the Romans generally exhibited exotic animals in venues such as the Colosseum as prizes from conquered foe in various regions of North Africa. Each Roman emperor had a collection of animals. The entertainment of the day pitted many of the animals against one another in staged fights, or the animals were slaughtered in gladiatorial sport, as opposed to being revered as sacred or used for educational purposes (Blair, 1929; Bostock, 1993; Hoage et al., 1996).

Maximilian II, the Holy Roman Emperor from 1564 to 1576, founded three zoos, which have the most direct link to the modern zoo. These institutions, at Ebersdorf (1552), Prague (1558), and Näugebäu (1558), all began with modest collections, which included wild horses, large felines, elephants, and ungulates. As time progressed, so did their collections and in 1736, Näugebäu received numerous animals from a private zoo at Belvédère after the death of its founder-owner. Soon thereafter, in 1752, much of the Näugebäu-Belvédère collection and stock moved to a new zoo in Austria at Schönbrunn, originally founded as a deer garden by Maximilian in 1569. Holy Roman Emperor Francis I presented the zoo as a gift to his wife Maria Theresa, and numerous expeditions to Africa and the Americas provided a lush variety of wildlife in its collections. In 1765, Joseph II, the son and successor to Francis I, opened Schönbrunn to the public, "for the benefit of the public, the glory of the gardens, and the advancement of science" (Fisher, 1966, p. 52). He also sent two collectors to the Americas in 1785, who returned with 12 "new" species of mammals and 250 species of birds. The Schönbrunn Zoo still operates today in its original function as the oldest surviving zoo and retains much of its eighteenth-century design and architecture; it is recognized as the first "modern zoo" (Fisher, 1966; Hoage et al., 1996; Kilfeather, 2005). The zoos described here descended from royal collections. The London Zoo at Regent's Park was the first zoo founded for general visitors to view exotic specimens.

In August 1824, under the leadership of Sir Thomas Stamford Raffles, a meeting of 21 "friends of a proposed zoological society" met in London. The "society" claimed that although Great Britain was much wealthier than any other country, it could not rival the "magnificent institutions," such as those in France, Sweden, and Vienna, in the exhibition of exotic plants and animals (Ritvo, 1996). The members of this committee were all either wealthy or a member of the aristocracy, and most of them were prominent scientists of that time. For example, Sir Humphrey Davy, President of the Royal Society and inventor of the miner's safety lamp, and Adolphus Seymour, 11th Duke of Somerset and eventual President of the Linnaean Society of London, were committee members. Officially, this group founded the Zoological Society of London (ZSL) in 1826, with Raffles as the first President, who died shortly thereafter. The Marquis of Lansdowne succeeded Raffles and was influential in obtaining the Royal Charter. The objectives of the ZSL were "the advancement of zoology and animal physiology, and the introduction of new and curious subjects of the animal kingdom" (Fisher, 1966, p. 5). In 1828, the Regent's Park Zoo (London Zoological Gardens, London Zoo) opened its gates only to members of the society and their guests, giving rise to the world's first scientific zoo and leading the way for the establishment of numerous zoological institutions and societies throughout Europe. In 1847, the zoo realized the financial value and public interest in its animals and opened its gates to the paying public to generate funds for animal care, as well as for scientific endeavors; the zoo quickly became a forum for social and recreational engagements (Turley, 1999).

A few years following the founding of the ZSL and the London Zoo, the Duke of Leinster called a meeting at the Rotunda Hospital in Dublin, and on 10 May 1830, issued a decree to form the Zoological Society of Dublin and to form a collection of animals based on the ZSL plan. The next year, Phoenix Park donated five and a half acres for the establishment of a zoological garden, which opened in September 1831 and is the third oldest public zoo in the world. The ZSL provided the majority of the original collection, and in the first year, the Dublin Zoological Gardens had over 30,000 visitors. To commemorate Queen Victoria's coronation, the zoo held an open (free) day in which approximately 20,000 people visited the zoo, still the highest number of visitors in a single day. Much history and myth concerning the Dublin Zoo circulate on the internet, most notably about a family who lost their son and eventually found him "traumatized" and visibly troubled in the penguin exhibit. When the mother got him home and ran him a bath, she found him playing in the tub with a live penguin chick, which he had smuggled out of the zoo in a backpack (Kilfeather, 2005). Presently, the Dublin Zoo enjoys much success and popularity as an "education first" institution. In 2010, 963,053 visitors passed through the gates, around 30,000 more than in 2009 (Dublin Zoo, 2011). This is a substantial number of visitors given that the total population of Ireland is around 4.5 million. In comparison, the Berlin Zoo had about 3.9 million visitors in 2010, with the total population in Germany being about 81 million (Blaszkiewitz, 2011).

Zoos in the United States of America

In the USA, the early Americans endured heavy hardships and were not overly concerned with a collection of wild animals. They had, however, a natural curiosity about the wilderness and the animals residing there. During this time, travelers would display a bear or other wild animal in a local tavern, take up a collection, and then move on to the next town. The size of these "traveling animal shows" increased, and American colonial merchants began to return from afar with exotic animals to sell for profits (Kisling, 1996). During the eighteenth century, local fairs exposed many people to exotic animals. Traveling showmen gave the public the chance to see wild animals and often taught the animals to perform entertaining tricks. The first traveling animals were an elephant, a rhinoceros, monkeys, and lions. In the United States, traveling shows and menageries flourished by 1813 (Baratay & Hardouin-Fugier, 2002). In 1841, Joel R. Poinsett addressed the National Institution for the Promotion of Science with a call for a national zoological park. Plans existed to develop the Smithsonian Institution in Washington, DC, but did not include the addition of a collection of live animals when the park opened in 1846. In March 1859, a meeting of naturalists and citizens in the Philadelphia home of Dr. William Camac, MD, led to the founding of the Zoological Society of Philadelphia. Dr. Camac had traveled extensively throughout Europe and felt that the inclusion of a zoological park would be a valuable addition to the community. The society laid plans for the construction of what would become the Philadelphia Zoological

Garden. The onset of the Civil War in 1861 put the plans for the institution on hold. In 1872, work began at Fairmount Park with the erection of a fence. An animal collector and the zoo's first superintendent, Frank J. Thompson, received permission to obtain the living collection. The founders wanted to create a unique atmosphere to set the Philadelphia Zoo apart from the traveling menageries and circuses that toured the country during the nineteenth century. The Philadelphia Zoological Gardens opened its gates to the public on July 1, 1874, with a collection of approximately 200 mammals on display, in addition to about 675 birds and 8 reptiles. Modeled after the London Zoological Garden, the Philadelphia Zoo employed a full-time staff and was the first exhibit in the country to house animals in permanent buildings. In 1901, Philadelphia also opened the first research laboratory associated with a zoo, the Penrose Research Laboratory (Hanson, 2002; Kisling, 1996). Some argue over which was actually the first American Zoo, some sources naming the Central Park Zoo as being in existence as early as 1859, although it was thought to be acting still as a menagerie and as a haven for abandoned circus animals (Blair, 1929; Croke, 1997; Fisher, 1966; Hanson, 2002; Horowitz, 1996; Kisling, 1996).

The Evolution of Zoo Design

During the mid-eighteenth century, zoo designs focused on taxonomy or the physical characteristics and relationships of the animals (Harrison, 1991; Hoage & Deiss, 1996; Karkaria & Karkaria, 1998; Rabb, 1994). Starting here, interpretation was restricted to the name of the animal and comparative anatomy and physiology (Karkaria & Karkaria). Zoos did little to promote the understanding of ecological relationships and importance of habitats. As zoos became a public institution in the nineteenth century, it became imperative that zoos spend time deciding how to best share the zoos' collections with visitors (Croke, 1997; Koebner, 1994). The need to provide visitors a place for public viewing gave rise to exhibits. A zoo exhibit is a space in which one or more animal specimens form the focal point (Tunnicliffe, 1996). Toward the end of the nineteenth century, scientific study of animal species found in zoos became more popular (Harrison).

Designers of the twentieth-century zoological park, or living museum, intend to show within an enclosure the natural habitat of the animal, introducing ecological themes and conveying information about the habitats and the behavioral biology of the animals. Exhibit designers often created moated enclosures meant to be similar to the animal's natural habitat (Baratay & Hardouin-Fugier, 2002; Croke, 1997; Harrison, 1991; Hoage & Deiss, 1996; Koebner, 1994). The naturalistic enclosures consisted of artificial rockwork of varying quality, sparse vegetation, and larger areas for the animal. The view of the visitor became more important. Moreover, zoo designers arranged animals zoogeographically into regional settings (Baratay & Hardouin-Fugier; Croke; Harrison; Hoage & Deiss; Koebner). However, the emphasis was on the exhibit itself, not science or education (Karkaria & Karkaria, 1998). The Modernism era followed the Zoogeographic Arrangement era. Health and sanitation became important issues, and designers converted animal enclosures to glass-fronted, laboratory-like enclosures. Rockwork of the previous era changed to plain walls of concrete and masonry. Interpretive messages of this era tended to be expanded labels, with information presented in lists or outlines, and placed outside the reach of the visitor and signs placed within the enclosures (Baratay & Hardouin-Fugier; Croke; Harrison; Hoage & Deiss; Koebner).

In the Postmodern Era, architectural style and graphics are sleek and futuristic. Computer graphic style signage competes with the live exhibits for attention. Animal shows also are part of the Postmodern Zoo (Hoage & Deiss, 1996; Koebner, 1994). Interpretive messages are restricted to animal behavior and other fascinating or entertaining facts (Baratay & Hardouin-Fugier, 2002; Croke, 1997; Harrison, 1991). Zoos in the twenty-first century still are evolving into conservation centers and places of conservation learning. The exhibit style associated with the future zoo is habitat immersion, which surrounds visitors with artifacts and educational stimuli directed toward a particular subject. In an immersion, exhibit there is a central theme, such as an ecosystem, region, or country, and everything in the exhibit orients to that theme. Thematic elements contribute to an integrated picture of the organism's habitat (Karkaria & Karkaria, 1998). An immersion exhibit is an enclosure design in which the visitor feels they are a part of the enclosure. The enclosure is landscaped with the use of both real and artificial material, giving the visitor an impression of the animal's real habitat (Kleiman, Thompson, & Baer, 2010). The interpretive design at such exhibits is unobtrusive (Baratay & Hardouin-Fugier; Croke; Harrison; Hoage & Deiss; Koebner). This is important, as the first goal of interpretation is to present an animal in relation to its habitat.

Today's zoo competes against digital entertainment, such as television, video games, and simulated environmental interactions. Therefore, zoos work to immerse the visitor in a sensory-laden multimedia, physical, and emotional experience. Zoos are developing interactive, immersion exhibits, such as the Bronx Zoo's Congo exhibit, Disney's Animal Kingdom, Henry Doorly Zoo, Cornwall's Eden Project, Zoo Zurich's Masoala Rainforest, Burger's Zoo, Singapore Zoo, Tennoji Zoo, and Polar Bear Conservation and Education Habitat. To further immerse visitors in the experience, increase visitation, and raise funds, zoos push the interaction of the visitor with the animals, such as the following: (1) The Alaska Zoo offers Keeper-for-a-Day, where a visitor spends the day with a zookeeper. They prepare food and feed the polar bears. (2) The San Antonio Zoo and Houston Zoo permits visitors to stand near the lions, while the handlers complete daily health checks. (3) The North Carolina Zoo invites visitors to feed the giraffes. (4) At the Melbourne Zoo, visitors may wash an elephant. (5) The Australia Zoo allows visitors to interact with the red panda.

In addition to the evolution of the purpose of zoos and their exhibit design, zoo art has progressed (Kleiman et al., 2010). In the early 1900s, zoos were built in a temple-like style that was meant to display the majesty and beauty of the animals. Zoos focused on animals as God's creations and strove to clearly depict their anatomy in artwork, such as animal murals on the back walls of cages. In the 1930s, this preference changed to more democratic illustrations of animals in relation to visitors and reflected the artistic tastes of popular culture. At the London Zoo, the architectural design of the animal exhibits tested new theories of humanity and design and sought to restore the harmony between people and nature. Zoologists of the time believed that humans and animals were closely connected and that visitors to the zoo would recognize the primitive levels of human nature in the zoo animals (Anker, 2006). The 1940s–1960s brought a new awareness of zoo animal welfare and zoo art focused on concepts of harmony and peace as well as the animals' vulnerability to humans. The 1980s saw a change in policy that no longer allowed visitors to come into contact with the zoo animals. To counteract the hands-off nature of the new policies, zoos installed life-size animal statues that were fully accessible to the visitors. Made out of sturdy materials such as stone or bronze, these animal sculptures were popular zoo attractions and encouraged tactile exploration by visitors. Contact with animal art was much safer than visitor-animal interactions.

Postmodernism of the 1990s has changed zoo art and challenged visitors' conceptions of animals, sought to provoke and unsettle visitors, and redefine the relationship between humans and animals. At the 2011 AZA conference, Dan Wharton, Senior Vice President Conservation Science Emeritus, Chicago Zoological Society, stated that "animals are art in protoplasm." Animals are no longer depicted sentimentally, but provoke visitors to "… rethink [their] preconceptions about animals, move beyond [their] self-centered understanding of the world, and take some responsibility for the endangered status of the animal" (Donahue & Trump, 2007, p. 122). Current zoo art continues the Postmodernism trend, encouraging visitors to think twice before anthropomorphizing the zoo animals. The North Carolina Zoo has a diverse collection of art and offers visitors an art at the park guide. Although many visitors may be uncomfortable with the realism of zoo artwork, it conveys an important message about the true nature of the relationship between animals and humans.

> The importance of art in zoos has changed throughout history, and the art in zoos has explored the human relationship with the animal world in ways that the actual animal exhibits could not. Successive waves of animal artists have tried to capture the "truth" about animals and zoos as a space. At various times the truth about animals has shifted from their anatomy, to their energy, to their vulnerability to human predation. (Donahue & Trump, 2007, p. 9)

Today, zoos have shifted from displaying human-created art to selling animal-created art. Elephants and orangutans at the Houston Zoo create original paintings that are sold to raise money for wildlife conservation and natural disaster relief funds (Turner, 2010). The Saint Louis Zoo allows the penguins to walk through paint and then onto paper to design colorful prints. The Nashville Zoo holds an Annual Animal Art Auction of cougar paw prints and artwork completed by macaws, elephants, red pandas, and meerkats. The Cincinnati Zoo and Botanical Gardens offers rhinoceros and elephant drawings.

However, in 2011, art took on a new form. The most recent infusion of art into zoos has developed via the Language of Conservation Project. The Language of Conservation Project has turned poetry into visual art within an exhibit space at the Audubon Zoo, Brookfield Zoo, Central Park Zoo, Jacksonville Zoo and Gardens, Little Rock Zoo, and Milwaukee County Zoo (Wharton, 2011). For example, in the anaconda exhibit at the Milwaukee County Zoo, visitors will find the following poem: "And in the deeps of a great water/the giant anaconda lies/like the circle of the earth . . ." (Neruda, 1974, p. 63). Including poetry in the exhibit design builds an emotional bridge between the visitor and the zoo because the poetry persuades people to think about nature. Poems can focus biological conservation knowledge in a new way and provide zoos with a new avenue to achieving their biological conservation education goal (Fraser, 2011). The exhibit embedded poems seem to be shaping intergenerational conversations by providing an interface between children and adult caretakers. Some children read the poems out loud, caregivers read the poems to children, and adults and children have been seen discussing the poems. Moreover, when visitors are asked if they noticed the poems during their visit, they say "no." However, when probed, zoo visitors can remember words or lines from the poems (Davis, 2011).

Zoo Education

After New York's Bronx Zoo opened its education department in 1929 (Schwammer, 2001), the idea spread that zoos could educate the public (Baratay & Hardouin-Fugier, 2002). A greater interest in natural history education also emerged as contact with nonurban nature began to dwindle. The first major European zoos to develop educational programs were the London Zoo in 1958 and Frankfurt Zoo in 1960 (Schwammer). Now, nearly all zoos in the United States of America (USA), Europe, and Australia, including small zoos, have Education Departments (Heimlich, 1996; Walker, 1991).

Post and van Herk (2002) outlined the four evolutionary phases of education in zoos: Phase I: the Prehistoric Period, Phase II: the Island Educator, Phase III: the Recognized Educator, and Phase IV: the Utilized Educator. During the Prehistoric Period, zoos did not have education departments or an educational influence, and curators designed the primary educational feature, signage. From 1960 to 1980, zoos identified the Island Educator as a misfit, tolerated by colleagues and isolated from the rest of the zoo. In the 1980s, the Recognized Educator was uninvolved in zoo design, but expected to implement the biological conservation message. Education was evolving into an important part of the mission of the zoo. Today, the Utilized Educator is recognized as important and has an active voice in the design of new exhibits. Education departments are evolving into designers of informative, educational signage and educational oases for school children.

Whereas zoological gardens enabled the public to view exotic, unfamiliar, and strange animals, the establishment of children's zoos had the opposite aim, to enable children to see familiar domestic animals and perhaps capture lost childhood experiences for adults. The first children's zoo in the world was in London at the Zoological Society's animal collection in Regent's Park, constructed to resemble a farm, providing domestic animals for children to view and pet. Over the years, the rationale for the London children's zoo changed, and various exotics joined the collection. In the early 1990s, the London Zoo demolished the old children's zoo and built a new one, very similar to the original.

The theme of the first US children's zoos (Everly, 1975) was nursery rhymes and fairy tales and utilized animals from these rhymes and tales. La Fontaine Park (zoo no longer open) in Montreal, Canada, was design based on the characters of Aesop's fables (Everly). Other more recent, purpose-built children's zoos, such as the Folsom Children's Zoo and Botanical Gardens in Lincoln, Nebraska, or Drusillas Children's Zoo at Alfriston, Sussex, England, orient their designs to assist children in developing their biological knowledge in a meaningful manner by building on their existing knowledge. These collections have domesticated animals and representatives of the major taxonomic groups. Interpretation and opportunities for interaction and learning are designed to match the abilities and conceptual development of children.

Modern zoos were founded for "the exhibition of curios" (Zuckerman, 1976) as well as for scientific research. Formal education at zoos became established in the second half of the twentieth century. In 1969 William G. Conway, a leading US zoo director, wrote in *Zoos: Their Changing Roles* that urban children have a need to be near live wild animals. The majority of zoos today are in urban locations or near the urban populace. Conway's vision was that urban zoos would fill an increasingly important role in education as well as in zoological conservation and research. This vision is important for zoos because much of what people learn occurs outside of the formal educational setting (Falk & Dierking, 2010; Patrick & Tunnicliffe, 2011; Tough, 1979; Tunnicliffe, 2010).

The Zoo Voice Today

E. O. Wilson (1993) stated that zoos must do three things to protect biodiversity: (1) educate, (2) argue, and (3) explain. Zoos are a place where humans come in contact with nonurban nature. Loss of human/nature contact leads to decline of knowledge of the natural world. Now is the time for zoos to motivate the members, the visitors, and the public to become involved in preserving nature (Nichols, 1996). Even though a living animal collection attracts visitors, education is not automatic (Walker, 1991). Therefore, zoos must consider their future and how it links to the public's understanding of biological concepts, conservation biology, and environmental change. Not only are zoos and aquariums caretakers of life in an ongoing extinction crisis (Kellert & Wilson, 1993), they are a gateway to conservation biology and biological conservation education.

Biology is the study of living things, including *Homo sapiens*, and their taxonomy, physiology, behavior, environmental needs, and interrelationships. Conservation in biological terms is defined as preventing the eradication of species, habitats, and ecosystems. The main principles, concepts, goals, and values of *conservation literacy* have been set out by the Society of Conservation Biology (Trombulak et al., 2004). The Society of Conservation Biology states that the central values of conservation biology that people should possess are represented by *conservation literacy*. Conservation literacy should be part of good citizenship and should be addressed by practitioners to make the public conservation-literate citizens. Biology conservation education (Tunnicliffe, 2010) is a subset of biology conservation or science education. Biology conservation education teaches the learner that the living world is of personal worth to themselves and their existence. Conservation biological science and education and the way of life being advocated are of value to the learner only if the learner views them as valuable. Therefore, unless the teaching of biological conservation genuinely engages learners and their real concerns, learners may acquire little understanding of biological conservation and not alter their attitudes or behaviors.

The Buffon Declaration (Buffon Symposium, 2007) recognizes that natural history institutions and science have a responsibility to maintain global biodiversity, sustainable management of biodiversity and ecosystems, and ultimately survival of the human population. Natural history institutions and zoos are important in that they (1) are primary repositories for scientific samples that provide an understanding of the variety of life, (2) provide research that analyzes the mechanisms of biodiversity, (3) offer education programs and expertise about how to tackle current and future environmental challenges, and (4) are cultural forums for direct engagement with society. The interactions between zoos, museums, and visitors are essential in provoking the changes in human behavior needed to preserve biodiversity and maintain a sustainable planet. Conservation biology education is the key to the initiatives of zoos and museums.

In addition to being fun play areas, children's zoos should be developed based on children's everyday experiences. Children have first-hand experiences with organisms in their local communities and acquire second-hand experiences from the media, internet, and books. Children do recognize organisms, specifically those organisms that have characteristics similar to humans. A children's zoo should have an area in which children are introduced to organisms and familiarized with the main collection of the zoo. An ideal children's zoo would relate an organism's features to a human (i.e., a child) by providing the child with help identifying the organism's life processes (feeding, behavior, excretion, etc.) and life needs (e.g., food gathering, habitat). Moreover, by using organisms to illustrate the range in which the biological functions/needs are met, the relationship between plants and animals could be explored, and the key role of plants could be emphasized. The ethical treatment of animals may be emphasized and addressed by discussing children's pets. In the 1990s, the London Children's Zoo focused on the ethical treatment of pet animals in the Pet Care Center. A more recent example of zoos incorporating personal pets into the education of children is taking place at the Hamill Family Play Zoo in Chicago, where zoo educators integrate the welfare of pets into their animal program.

Today, the Zoo Voice focuses on zoological taxonomy and conservation of the natural world. The zoo ethos is conservation and now includes conservation of endangered species (plant and animal), oceans, and land spaces. Additionally, zoos have begun to focus on issues of climate change, human/natural world interactions, and human impact on the natural world. Increasingly, the Zoo Voice is stating that visitors should not only be aware of the animals but should relate the animals to environmental issues, ethics, and concerns. Even though the main focus of visitors is the animals on display, the Visitor Voice in this book is described as the content of the dialogues that take place during a zoo visit. We look at the bidirectional conversations employed by visitors and facilitators.

As zoos provide interactive, play areas in addition to children's zoos, they provide out-of-door activities that allow children to interact with each other, adults, and the natural environment. The Brookfield Zoo has applied the findings of conservation psychology in its design of the Hamill Family Play Zoo (Saunders, 2003). The Brookfield Zoo hopes to develop an emotional association between visitors and the natural world and to motivate visitors to take conservation action. Additionally, by focusing on conservation psychology, the zoo's aim is to promote conservation of the natural world in its widest sense, not just conservation of species, but understanding the intricate web of biological diversity. Conservation biology education is developing to compliment the advances in conservation psychology (Tunnicliffe, 2010). The educational relationships zoos build with the public are a perfect way for zoos to share their biological conservation message. Understanding the historical development of zoos is an important aspect in defining their educational future.

References

Anker, P. (2006). Bauhaus at the zoo: Modernist designers in the 1930s found inspiration in the life sciences. *Nature, 439*, 916.

Ballantyne, R., Packer, J., Hughes, K., & Dierking, L. (2007, July). Conservation learning in wildlife tourism settings: Lessons from research in zoos and aquariums. *Environmental Education Research,13*(3), 367–383.

Baratay, E., & Hardouin-Fugier, E. (2002). *Zoo: A history of zoological gardens in the West.* London: Reaktion Books, Ltd.

Blair, W. (1929). *In the zoo.* New York/London: Charles Scribner's Sons.

Blaszkiewitz, B. (2011, January 25). *Animal statistics 2010.* Retrieved from www.zoo-berlin.de/zoo/tiere-wissenswertes/tierstatistik/tierstatistik-2010.html

Bostock, S. S. (1993). *Zoos and animal rights: The ethics of keeping animals.* London: Routledge.

Buffon Symposium. (2007, October 18–19). *The Buffon Declaration: Natural History Institutions and the Environmental Crisis.* Concluding message from the Muséum National d'Histoire Naturelle, Paris.. Retrieved from http://www.bfn.de/fileadmin/ABS/documents/BuffonDeclarationFinal%5B1%5D.pdf

Conway, W. G. (1969). Zoos: Their changing roles. *Science, 163*, 48–52.

Croke, V. (1997). *The modern ark: The story of zoos, past, present and future.* New York: Scribner.

Davis, B. (2011, September 12–17). Educating the visitor with poetry. *Poetry and the Wild: The Language of Conservation Project.* Presentation at the Association of Zoos and Aquariums Conference, Atlanta, GA.

Dembeck, H. (1965). *Animals and men.* New York: Natural History Press.

Dick, G. (2010, September 3–8). *Evolution of zoos.* Paper presented at the Association of Zoos and Aquariums Conference, Houston, TX.

Donahue, J., & Trump, E. (2007). *Political animals: Public art in American zoos and aquariums.* Lanham, MD: Lexington Books.

Dublin Zoo. (2011, January 25). *A record breaking year at Dublin Zoo.* Retrieved: www.dublinzoo. ie

Everly, R. E. (1975). *Fun, fantasy and function in children's zoos.* First International Symposium on Zoo Design and Construction, Paignton Zoo, UK.

Falk, J., & Dierking, L. (2010). The 95 percent solution: School is not where most Americans learn most of their science. *American Scientist, 98,* 486–493.

Falk, J. H., Reinhard, E. M., Vernon, C. L., Bronnenkant, K., Heimlich, J. E., & Deans, N. L. (2007). *Why zoos and aquariums matter: Assessing the impact of a visit to a zoo or aquarium.* Silver Spring, MD: Association of Zoos and Aquariums.

Fisher, J. (1966). *Zoos of the world.* London: Aldus Books.

Fraser, J. (2011, September 12–17). Design of public poetry installations and the results of impact studies. *Poetry and the Wild: The Language of Conservation Project.* Presentation at the Association of Zoos and Aquariums Conference, Atlanta, GA.

Gray, B. (1991). *The San Diego Zoo.* San Diego, CA: Zoological Society of San Diego.

Hancocks, D. (2001). *A different nature: The paradoxical world of zoos and their uncertain future.* Los Angeles: University of California Press.

Hanson, E. (2002). *Animal attractions: Nature on display in American zoos.* Princeton, NJ: Princeton University Press.

Harrison, B. (1991). *The future evolution of zoos.* Paper presented at the conference of the International Union of Directors of Zoological Gardens, Singapore.

Heimlich, J. (1996). *Adult learning in nonformal institutions.* Eric Digest, ED399412, 1–7.

Hoage, R., & Deiss, W. (1996). *New worlds, new animals: From menagerie to zoological park in the nineteenth century.* London: The Johns Hopkins University Press.

Hoage, R., Roskell, A., & Mansour, J. (1996). Menageries and zoos to 1900. In R. Hoage & W. A. Deiss (Eds.), *New worlds, new animals: From menagerie to zoological park in the nineteenth century* (pp. 8–18). London: Johns Hopkins University Press.

Horowitz, H. L. (1996). The national zoological park: "City of Refuge" or zoo? In R. Hoage & W. A. Deiss (Eds.), *New world, new animals: From menagerie to zoological park in the nineteenth century* (pp. 126–135). London: The Johns Hopkinks University Press.

Karkaria, D., & Karkaria, H. (1998). Zoorassic Park: A brief history of zoo interpretation. *Zoos' Print, 14*(1), 4–10.

Kellert, S., & Wilson, E. (1993). *The biophilia hypothesis.* Washington, DC: Island Press.

Kilfeather, S. (2005). *Dublin: A cultural history.* New York: Oxford University Press.

Kisling, V. N. (1996). The origin and development of American zoological parks to 1899. In R. Hoage & W. A. Deiss (Eds.), *New world, new animals: From menagerie to zoological park in the Nineteenth Century* (pp. 109–125). London: The Johns Hopkins University Press.

Kleiman, D., Thompson, K., & Baer, C. (2010). *Wild mammals in captivity: Principles & techniques for zoo management.* Chicago: The University of Chicago Press.

Koebner, L. (1994). *Zoo book: The evolution of wildlife conservation centers.* New York: Tom Doherty Associates.

Neruda, P. (1974). *Five decades: Poems 1925–1970 (Neruda, Pablo).* Buenos Aires, Brazil: Grove.

Nichols, M. (1996). *Keepers of the kingdom: The new American zoo.* Richmond, VA: Thomasson-Grant & Lickle.

Patrick, P., & Tunnicliffe, S. (2011). What plants and animals do early childhood and primary students' name? Where do they see them? *Journal of Science Education and Technology.* Available at: http://www.springerlink.com/content/e27121057mqr8542/

Post, H., & van Herk, R. (2002). Education and exhibit design. *EAZA News, 37,* 10–12.

Rabb, G. (1994). The changing roles of zoological parks in conserving biological diversity. *American Zoologist, 34,* 159–164.

Ritvo, H. (1996). The order of nature: Constructing the collections of Victorian zoos. In R. Hoage & W. A. Deiss (Eds.), *New worlds, new animals: From menagerie to zoological park in the nineteenth century* (pp. 43–50). London: The Johns Hopkins University Press.

Rose, M. (2010). World's first Zoo-Hierakonpolis, Egypt. *Archaeology, 63*(1), 25–32.

Saunders, C. (2003). The emerging field of conservation psychology. *Human Ecology Review,* *10*(2), 137–149.

Schwammer, G. (2001). Education: On-site programs. In C. Bell (Ed.), *Encyclopedia of the world's zoos.* Chicago: Fitzroy.

Tough, A. (1979). *The adult's learning projects.* Toronto, ON: Ontario Institute for Studies in Education.

Trombulak, S., Omland, K., Robinson, J., Lusk, J. Fleischner, T., Brown, G., & Domroese, M. (2004). Principles of conservation biology: Recommended guidelines for conservation literacy from the education committee of the *Society for Conservation Biology, 18*(5), 1180–1190.

Tunnicliffe, S. D. (1996). Conversations with primary school parties visiting animal specimens in a museum and zoo. *Journal of Biological Education, 30*(2), 130–141.

Tunnicliffe, S. D. (2010, September 2–8). *What is conservation biology education (CBE)?* Poster presented at the the Association of Zoos and Aquariums Conference, Houston, TX.

Turley, S. K. (1999). Conservation and tourism in the traditional UK zoo. *The Journal of Tourism Studies, 10*(2), 2–13.

Turley, S. K. (2001). Children and demand for recreational experiences: The case of zoos. *Leisure Studies, 20*(1), 1–18.

Turner, A. (2010, March 10). Abstract animal art may mean aid to Haiti. *Houston Chronicle.* Retrieved from http://www.chron.com/disp/story.mpl/metropolitan/6907358.html

Wagoner, B., & Jensen, E. (2010). Science learning at the zoo: evaluating children's developing understanding of animals and their habitats. *Psychology & Society, 3*(1), 65–76.

Walker, S. (1991). *Education and training in captive animal management.* Proceedings in Perspectives in Zoo Management, National Zoological Park, New Delhi Zoo Ed Book, Zoo Outreach Organization.

Wharton, D. (2011, September 12–17). Creating a vision for the language of conservation. *Poetry and the wild: The language of conservation project.* Presentation at the Association of Zoos and Aquariums Conference, Atlanta, GA.

Wilson, E. O. (1993). *The diversity of life.* Cambridge, MA: Harvard University Press.

Zuckerman, S. (1976). *The zoological society of London 1826–1976 and beyond.* London: Academic.

Chapter 3
Rationale for the Existence of Zoos

Ultimately, in this century, we have begun to realize that the way we display animals truly affects the way people view them and people speaking reverentially, and quietly before exhibits that were truly natural. Just as often standing before, old barren cages, I saw zoogoers yell, throw food and make fun of the animals inside. (Croke, 1997, p.93)

Today's zoos serve two basic functions: community resource and conservation entity (Hanna, 1996, p. 76).

Some individuals and groups view zoos and aquariums as prisons for animals. If we, as zookeepers, maintain our facilities with humanity and high standards, we needn't hang our heads and call ourselves wardens. We can instead look at our facilities with pride and see them as bridges between our visiting public and the wild they may never see. We can look our captive charges in the eye as we treat them with the respect due the highest-level ambassadors—ambassadors of the wild. (Hanna, 1996, p. 82)

Today, zoos, by their own definition, are conservation organizations and recognize themselves as places which are involved in the conservation of flora/fauna. However, the primary reasons cited by visitors for visiting a zoo are for the educational benefit of children and to see animals. Animals *can* and *do* interest visitors without the additional interpretation provided by institutions. As a specimen on display in a zoo, the animal becomes an "exhibit" and takes on the "mantle of history" and becomes part of the story that the zoo wants to tell. The animal specimen is part of the evidence for its species characteristics, just as a human artifact is evidence for aspects of human civilization. Furthermore, industrialization and urbanization are reducing students' direct interactions with nonurban nature. Due to the reduced contact people have with nonurban nature, interest in the variety of living things is perhaps becoming redirected toward human artifacts. As the world becomes more urbanized, our personal experiences with animals become more isolated, in many cases limited to domesticated pets and urban species. No matter how zoos choose to get their message to visitors, education is their most important conservation function.

Today, more than ever, zoos need to think harder [about] why they are there and what role they will fill in conservation, education, and research. Millions of dollars go to house artwork in museums, but there are more Rembrandts in the world than there are Siberian tigers. (Hutchins, 2003, p. 25)

P.G. Patrick and S.D. Tunnicliffe, *Zoo Talk*, DOI 10.1007/978-94-007-4863-7_3,

As discussed in the second chapter, the functionality of zoos has evolved drastically since the 1820s, when the main focus was to display a vast collection of exotic animals for public enjoyment (Conway, 2003; Rabb, 2004). Zoos originally emerged as a place of scientific collections. The London Zoo is the predecessor of the modern zoo. This institution's founding body, the Zoological Society of London (ZSL), envisioned the zoo as a scientific collection for the advancement of zoology as well as comparative physiology of animals. In the beginning, the zoo was open only to members of the ZSL and their guests, but the financial demands of maintenance and upkeep influenced the opening of the zoo to a paying public. The business of the zoo would from then on be influenced by the zoo visitor, and the survival of zoos would be based upon the public's perception of the everyday role it plays in society.

Originally, most zoos were created as a place for recreation with an emphasis on biological literacy. The conservation of wildlife diversity and biological conservation education were not the most pressing issues to zoo founders. Indeed, the original pedagogical approach of zoos was to display animals in rows of enclosures so that people could see strange creatures and make comparative observations of the physical form of different species. As zoos developed during the 1960s, in addition to places of fun and family recreation, they saw themselves as having four functions: recreation, education, research, and conservation (Nichols, 1996). In the 1970s, as ecological concerns began to emerge, zoos could no longer justify themselves as primarily entertainment facilities and started to consider making conservation their central role (Hancocks, 2001). During the 1970s, zoo professionals began conservation programs, and the American Association of Zoos and Aquariums (AZA) maintained that conservation had become its highest priority (Hancocks, 2001; Reed, 1973). Today, due to the deteriorating link between humans and the nonurban natural world, zoos are coming under pressure to develop conservation plans and educate the public about the living world while maintaining financial stability. Therefore, zoos have evolved to include education as a priority along with conservation and research (Karkaria & Karkaria, 1998; Patrick, Matthews, Ayers, & Tunnicliffe, 2007a; Patrick, Matthews, Tunnicliffe, & Ayers, 2007b). Modern zoos work to bring biological conservation to the forefront of their educational programs and have the potential to shape public opinion, to encourage empathetic attitudes toward wildlife, and to educate the public about ecology, evolution, and wild organisms.

Presently, zoos must justify their existence against a constant barrage of anticaptivity and animal rights groups, who dispute that any education is taking place and promote the idea that zoos exist purely as a form of entertainment at the expense of the organism's natural instincts. Some critics of zoos have gone so far as to compare the exhibits of animals to pornography and suggest there is a negative educational impact on zoo visitors (Acampora, 2005; Wagoner & Jensen, 2010). Acampora (1998) believes that zoos are not educationally beneficial because

> ...the public is largely indifferent to zoo education efforts (few stop even to look at, let alone read, explanatory placards); animals are viewed briefly and in rapid succession; people tend to concentrate on so-called babies and beggars—their cute countenances and

funny antics capture audience attention (Ludwig 1981). Of course, this sort of amusement is at the heart of what a zoo is (scientific ideologies of self-promotion notwithstanding). Consequently, and insidiously, what visits to the zoo instruct and reinforce over and over again is the subliminal message that nonhuman animals are here in order to entertain us humans. Even when, during our deluded moments of enlightenment, we insist that they are here rather to edify—even then their presence is still essentially assigned *to or for us*. Thus the phenomenological grammar of their appearance precludes the possibility of full otherness arising; this is what it means to put and keep a live body on display (a structural inauthenticity [sic] that remains despite the best intentions of humanitarian/ecologic pedagogy). (p.2)

The evolution of how a zoo defines itself is dependent on the public's perceptions of the zoo's contributions to society. The problem is that the day-to-day operations of the zoo go unnoticed, and citizens do not realize that the zoo's practices have an immeasurable and direct impact on the community and wildlife. Therefore, it is important to validate the existence of zoos through the programs they offer, their contributions to society, and the perceptions that the zoo-going public hold regarding these institutions. Zoos typically identify their own self-worth through their mission statement.

The [AZA] (2011a) defines the mission of zoos as striving to be global leaders in "animal care, wildlife conservation and science, conservation education, the guest experience, and community engagement." Patrick, Matthews, Ayers & Tunnicliffe, (2007a) and Patrick, P., Matthews, C., Tunnicliffe & Ayers, (2007b) analyzed AZA-accredited zoos' mission statements and established that there are seven main themes in zoo mission statements: (1) *education*, (2) *conservation*, (3) *recreation*, (4) *facilities*, (5) *research*, (6) *administration*, and (7) *culture*. In addition to the seven themes found in mission statements, the literature identifies five main purposes of zoos: (1) exhibiting animals for the public (Mazur & Clark, 2001), (2) providing education (Clayton, Fraser, & Saunders, 2009; Ogden & Heimlich, 2009; Owen, Murphy, & Parsons, 2009; Price, Vining, & Saunders, 2009; Visscher, Snider, & Stoep, 2009), (3) conservation (Ballantyne, Packer, Hughes, & Dierking, 2007; Wagoner, Chessler, York, & Raynor, 2009), (4) research (Fernandez & Timberlake, 2008; Hutchins & Thompson, 2008; Kolbert, 1995; Rabb, 2004), and (5) providing recreational opportunities for visitors (Bostock, 1993; Chizar, Murphy, & Lliff, 1990; Martin, 2000). There are overlaps in how zoos define themselves through their mission statements and how the literature defines the purposes of zoos. Even though modern zoos are placed in a unique position in which they can serve numerous functions, the average zoogoer does not understand the intricacies and mélange of their purposes (addressed in Chap. 4).

Education

Part of realizing the zoo's role in conservation is informing and involving the public in the conservation mission. Zoos are in a unique position to provide environmental education and biological conservation education to large numbers of visitors. In 2007, 132 (96%) of the 137 AZA-accredited zoo mission statements include education as a predominant theme (Patrick, Matthews, Ayers, et al., 2007; Patrick, Matthews, Tunnicliffe, et al., 2007). However, no mission statements made a direct or specific reference to biological conservation education. Mission statements used the words conservation and education but not together. Therefore, a reference to education in the mission statement does not match the literature's specific call for biological conservation education. As early as 1989, the Zoological Society of Philadelphia stated that the modern zoo was failing to realize its potential for biological conservation education. Moreover, in 1994, Koebner stated that biological conservation education had become the first priority of accredited zoos and aquariums. The 1993 Conservation Organization Strategy (IUDZG/CBSG, 1993) developed specific goals for zoo conservation education: "(1) make it clear that nature conservation affects everyone, (2) increase public awareness of the connections between consumption and lifestyle and the survival of species and bio-logical systems, and (3) inform the public about the threatened status of animals (p. 25)." However, zoo mission statements do not state they are specifically interested in conservation education. Zoo mission statements identify education as a general term, which could include knowledge of plants and animals, taxonomy, habitats, behavior, etc. Since 1994, the literature has focused specifically on biological conservation education (Clayton, Fraser, & Saunders, 2009; Ogden & Heimlich, 2009; Owen, Murphy, & Parsons, 2009; Price, Vining, & Saunders, 2009; Visscher, Snider, & Stoep, 2009) instead of the previously mentioned learning about plants and animals. Now, individual zoos need to specify biological conservation education as a mission and purpose, if biological conservation education is in fact their main goal.

Zoo mission statements may not define zoos as responsible for conservation education, but today's research literature specifically identifies conservation edu-cation as the prominent theme of zoos. A Google Scholar (http://scholar.google.com) search, inserting "zoo conservation education" as the search term, reveals that 5,500 documents are available on the topic since 2007. Therefore, zoo education research is shifting toward looking at zoos as sources of conservation education. The specific aim of biological conservation education is to develop lifelong knowledge and skills for conservation action. Biological conservation education recognizes the central role of people in all nature conservation efforts and is designed to people and their perceived roles in nature. Biological conservation education promotes the public education and awareness of the conservation of biodiversity by providing information about species and their natural habitats and working to develop a relationship between the public, wildlife, and wild habitats (AZA, 2011a; International Zoo Educators Association [IZEA], 2011; World Association of Zoos and Aquariums [WAZA], 2011). The zoos of the world have a unique role to play in the global efforts to make people conscious of the role of zoos in biological conservation (Carr, 2011; Hancocks, 2001; IUDZG/CBSG, 1993; IZEA, 2011; Sommer, 1972). Therefore, it is not only important that zoos formally define their role in education, but it is imperative that they define their role in biological conservation education.

The World Conservation Strategy (IUDZG/CBSG, 1993) defines conservation education in zoos as

> ... explaining the irreplaceable value of the entire biological system of our planet and all of its constituent components... conservation education in zoos can make it clear that nature conservation affects everyone, and that everyone needs to be concerned with it. They should play an active role in increasing the public and political awareness of the connections between consumption and lifestyle and the survival of species and biological systems. Conservation education includes informing the public about the threatened status of the species of zoo animals, and other animals which are taxonomically and/or geographically related. Conservation education in zoos can make it clear that nature conservation affects everyone, and that everyone needs to be concerned with it. (p. 25)

The International Zoo Educators Association states that biological conservation education is

> ... the process of influencing people's attitudes, emotions, knowledge, and behaviors about wildlife and wild places. This is done through the efforts of skilled educators and interpreters, who use a variety of techniques, methods, and assessments to reconnect people with the natural world. (IZEA, 2011)

Zoos give visitors the opportunity to see unique organisms, as well as a chance to learn more about familiar animals. During these close encounters, zoos have the attention of individuals, which gives them an opportunity, be it short, to emphasize their importance and the importance of conservation (Falk & Dierking, 2000; Tunnicliffe, 1995, 1996). Currently, conservation messages are shifting from endangerment and captive breeding to the importance of saving habitat (Mazur & Clark, 2001). As zoo programs continue to grow and improve, students and teachers remain a core audience served by classes, tours, outreach programs, and special curricular materials. Progress in conservation depends on reaching out to schools through educational opportunities and advancing public understanding of science issues and human roles in conservation.

Conservation

Conservation of endangered species and their habitats is a key role of zoos. Through their mission statements, zoos characterize their role in conservation as practice and/or advocacy (Patrick, Matthews, Ayers, et al., 2007a; Patrick, Matthews, Tunnicliffe, et al., 2007b). AZA defines zoos as conservation centers that are responsible for ecosystem health and species survival. To inform AZA of their conservation and research contributions, each AZA-accredited zoo submits data to the AZA Annual Report on Conservation and Science (ARCS) database. The ARCS database accounts for the $90 million that zoos spend each year on conservation initiatives and tracks over 3,700 conservation projects (AZA, 2011b). The World Association for Zoos and Aquariums has branded over 185 projects as conservation endeavors. Fifty percent of these projects were directed at the conservation of mammals (Dick, 2010). Zoos' conservation programs are important because the rate at which species are becoming extinct or threatened is occurring at a faster rate than at any time in Earth's recent history (Rogers, & Laffoley, 2011; Convention on Biological Diversity, 2010).

Zoological institutions are continuing to evolve into centers of wildlife conservation (Rabb & Saunders, 2005). Human population growth has led to the loss of habitat, habitat fragmentation, overhunting, climate change, and invasive species producing unnatural environmental stresses on wild populations (Bertram, 2004). The extinction of a single plant or animal has an unmeasurable effect on the surrounding ecosystem and represents not only the loss of a plant or animal but also the loss of millions of years of genetic evolution and adaptation (Cameron et al., 2011; Fonseca, 2009). In order to combat the loss of genetic variability, many zoos have frozen zoos, in which they freeze plant seeds and/or animal tissue. Plant-frozen zoos have been established by the Millennium Seed Bank Kew Gardens (UK) and the Missouri Botanical Garden (USA). Animal-frozen zoos were first established at the San Diego Zoo in the 1970s (Linington, 2000). In 2004, the Frozen Ark Project was established and invited zoos to be active participants. The Frozen Ark Project saves animal tissue with the goal of saving "the genetic material of threatened animals species and, where possible, their viable cells before they become extinct" (Clarke, 2009, p. 222). Each 1 cm^2 tissue contains thousands of cells and holds the entire genome of the animal. By sampling and preserving DNA, viable somatic cell cultures, and gametes in cryogenic labs, the Frozen Ark Project offers researchers the ability to develop new *in vitro* and *ex situ* techniques which can save some species from complete extinction, especially those which are difficult to breed in captivity (Clarke). These vast databases can lead to a better understanding of the interplay of cytogenetics and infectious disease and play a more crucial role in bio-surveillance. With the recent outbreaks of West Nile virus, SARS, and monkeypox, researchers have turned to zoos' frozen collections for genetic samples and tissue.

All AZA-accredited zoos are required to have a conservation plan in their mission statement, and the goal of each institution should be to have a "measureable impact on wildlife conservation" (AZA, 2011b). However, setting zoos' goals and policies for conservation action is increasingly a scientific, monetary, and political endeavor with numerous factors influencing the selection of ecosystem species (Leader-Williams et al., 2007). Some AZA-accredited zoos do participate in restoration programs. People believe that zoos are saving large numbers of mammalian species, when in reality, much of the conservation that occurs is the sustainability of small populations of animals. The conservation of these small numbers is safeguarding the species against extinction with the prospect of reintroducing them into the wild. Additionally, many of these species are used to raise funds and public awareness of environmental issues such as deforestation and habitat destruction (Bertram, 2004; Turley, 1999).

There are several examples of the conservation work zoos do with small, localized populations. One such example is the California Condor. In 1952, the San Diego Zoo established a captive breeding program to increase the wild populations.

In the 1980s, growing concern of high mortality rates for the California Condor led to a massive effort to rescue this majestic animal from sure extinction as a result of lead poisoning from ingesting bullets lodged in carcasses. By 1987, only 27 individuals remained, but breeding efforts saw an exponential growth of captive populations, and by 1992, reintroduction efforts began. By 1999, 88 birds had been released in 16 separate attempts (Meretsky, Snyder, Beissinger, Clendenen, & Wiley, 2000). Currently, the Los Angeles Zoo, the San Diego Wild Animal Park, the Portland Zoo, and the World Center for Birds of Prey in Idaho have very successful captive breeding programs. There are around 370 living California Condors with about 180 living in the wild, but the threats of lead poisoning and habitat destruction are still difficult barriers in the wild populations. Other US reintroduction programs include the black-footed ferret at the Cheyenne Mountain Zoo and the Smithsonian National Zoo, the Karner blue butterfly at the Toledo Zoo, and the desert antelope and the Mexican wolf also at the Smithsonian (AZA, 2011c; Smithsonian National Zoological Park [SNZP], 2011; Sweet, 2006).

One of the main tenets of zoo mission statements is the involvement of zoo staff and researchers in zoo wildlife conservation programs. However, zoos need to involve the public in their conservation and reintroduction programs to make them more successful. For example, the Durrell Wildlife Conservation Trust runs the Toadwatch campaign. The Toadwatch campaign asks people in the community to report when and where they see the Jersey toad (*Bufo bufo*), known locally as crapaud. Since 2005, the trust has recorded over 300 separate sightings and determined that the Jersey toad population is declining in natural ponds and increasing in garden ponds. In addition to reporting their Jersey toad sightings, volunteers participate in the Toads on Roads campaign. On wet, warm winter nights, volunteers pick up toads, weigh and measure them, and carry them across busy roads (Durrell Wildlife Conservation Trust, 2011). Zoos are also working with local farmers and local community environmental watch programs to develop conservation programs. For example, the Wildlife Conservation Society (WCS) is working with Indonesian farmers in southern Sumatra to develop a peaceful coexistence with local elephants (http://www.wcsip.org/). Moreover, the WCS has worked with the local community and conservation groups to bring back the Bronx River in New York. The Bronx River was heavily polluted but is now home to native fish species that have since returned.

The work at the Durrell Wildlife Conservation Trust proves that citizen scientists are an important resource in gathering large amounts of data over a vast area and *citizen conservationists* are an essential part of ensuring the survival of a species. *Citizen conservationists* are individuals who are inspired to take action in caring for natural resources, maintaining local wildlife habitats, and participating in local environmental conservation efforts. It is now the role of the zoo to figure out how their work in conservation and their animal displays can be a bridge between people and conservation action, inspiring individuals to become *citizen conservationist*.

Recreation or Entertainment

Even though zoos tout conservation and biological conservation education as their main goals and promote a conservation image, 69% of AZA-accredited zoo mission statements state that the zoo is for recreation (Patrick, Matthews, Ayers, et al., 2007a; Patrick, Matthews, Tunnicliffe, et al., 2007b). Moreover, public perceptions persist that zoos are places of entertainment rather than institutions of scholarly, scientific, or conservation pursuits (Bitgood, 1988; Frost, 2010; Kellert & Dunlap, 1989). Even though 60% of US zoo visitors state that zoos are places for education (Lessow, 1990), few people visit zoos with the declared aim to be educated. People generally visit zoos to be entertained (Martin, 2000). Zoos in developed countries compete with other attractions for the leisure time of visitors, while zoos in less-developed countries are one of the few available places for recreation. Even when adults recognize the educational importance of zoos, they do not visit the zoo intending to learn, but they encourage their children to do so (Lessow, 1990). As far back as 1885, the prospectus of the Zoological Society of London suggested that "vulgar admiration" was not the objective of their animal collection (Jordan & Ormrod, 1978). However, despite the educational and scientific aspirations of the learned society, the zoological garden became a place for a leisure visit.

Balancing the role of conservation advocate, educator, and recreational promoter is a major issue facing zoos (Tribe, 2004). However, when zoo specialists say that "you should use your selling experiences to advocate your mission (Holst, 2010)" and that zoo websites are "places to increase revenue (Israel, 2011)" and "the website's number one job is commerce not conservation education (Israel, 2011)," then the mission is entertainment not conservation. In the past, zoos have relied solely on live animals to draw visitors to the zoo. Today, the mission is selling an experience that the zoo hopes will bring people to the end goal of recognizing the importance of the zoo. In the past, zoos have utilized animal shows, 3D IMAX adventures, traveling art shows, informational carts, overnight camps, Halloween celebrations, and the announcement of babies on display to entertain people and increase visitorship. In today's digital world, these ploys are not enough. To address the need for more entertainment experiences, zoos provide enrichment activities for the visitor as well as for the animals. The Ocean Park zoo of Hong Kong and the Chessington World of Adventures in England combine animal exhibits with entertaining, rollercoaster rides. Visitors are provided with the opportunity to get their photo taken with an elephant, wash an elephant, participate in ropes courses, ride on zip lines, and ride in swan boats (Ellis, 2010; Garner, 2010; Holst, 2010). The John Ball Zoo recently added ropes courses to get 12–13-year-olds to visit the zoo. Holst stated that since the ropes courses were added, sales in the gift shop had increased, and zoo visitors stayed 20 minutes longer, thus increasing the zoos opportunity to deliver their biological conservation message to visitors. Providing zoo visitors with entertainment for an additional 20 minutes does not translate to more time to teach them the conservation message of the zoo, but 20 minutes of involvement in a conservation-related project does.

Facilities

Facilities have been addressed in 54% of AZA mission statements. Patrick, Matthews, Ayers, et al. (2007) and Patrick, Matthews, Tunnicliffe, et al. (2007) have found that the term most prominently used when describing the zoo's facilities is exhibits. The educational importance and influence of exhibits are discussed in Chap. 5; therefore, exhibits are not discussed in detail in this chapter. At present, zoos are changing their approach to exhibit design and their facilities. Hands-on laboratories, interactive technology, computer simulations, teleconferencing centers, discovery rooms, and immersion exhibits are now an integral part of what zoos have to offer. Classrooms are being built around themes such as habitats, biogeography, and educational expeditions. Zoos are taking into account their own learning objectives and that of national and state educational standards. Zoos are developing zoo schools in cooperation with their local school system. The idea of education for all visitors is apparent in their more recent pursuits to build libraries and teaching resource centers (Carr, 2011).

In addition to exhibiting animals, the institution's facilities must be maintained. The everyday maintenance of the zoo, including animal care, is a sensitive matter that requires knowledge and skill. The average visitor may think that animal care means that the animals are clean, properly fed, and have adequate space, but there are many facets of animal care. Superior health, husbandry, and welfare of zoo collections are of paramount importance to any AZA – (AZA, 2011b) and European Association of Zoos and Aquaria (EAZA, 2011a) – accredited institution. The priority of the zoo is not merely feeding the animals. Animal diet is an important aspect of maintaining healthy, breeding animals. EAZA (2011b) states that:

> Providing a good diet, which fulfills as many aspects of natural feeding ecology as possible, is an essential consideration in the welfare of zoo animals. As a basic foundation of animal management, nutrition is also integral to longevity, disease prevention, growth and reproduction. Proper feeding management incorporates husbandry skills and applied nutritional sciences.

During the 1960s, many zoos employed the services of licensed veterinarians to monitor the health and breeding of captive animals (Puan & Zakaria, 2007). The AZA recommends that all accredited institutions employ an on-site veterinarian. However, in some cases, zoos do not have the financial support to employ a full-time veterinarian. Therefore, all zoos are required to have a part-time veterinarian that inspects the collection twice a month and is able to respond quickly in case of emergencies. In other words, zoos must have a veterinarian available 24 hours a day. Additionally, zoos participate in year-round pest control, test for pathogens that could potentially ravage their collections, and use preventative measures such as quarantining newly arrived and sick animals (AZA, 2011c).

Research

AZA (2011b) states that
   A commitment to scientific research, both basic and applied, is a trademark of
the modern zoological park and aquarium. An AZA accredited institution must have a
demonstrated commitment to scientific research that is in proportion to the size and scope
of its facilities, staff and animal collections (p. 14).

Even though research is considered essential by zoological institutions
(Benirschke, 1987) and 37% of AZA-accredited zoos include research as a mandate
in their missions (Patrick, Matthews, Ayers, et al., 2007; Patrick, Matthews,
Tunnicliffe, et al., 2007), historically, zoos have not been regarded as elite
research institutions. Previously, zoos have worked with each other to share animal
collections and study behavior of captive animals. The *in situ* and *ex situ* research
conducted by zoos has traditionally been considered isolated from universities and
research institutions (Turley, 1999). Today, however, zoological institutions are
becoming more engaged in collaborative studies with major research institutions.
Moreover, the EAZA states that one of their main objectives is "to promote the
potential conservation value of zoo and aquarium research among authorities,
universities, and conservation bodies (EAZA, 2011c)." Zoos are involved in animal
research programs at universities and in research concerning infectious diseases
(McNamara, 2007; Turley, 1999).

Zoological institutions' *in situ* scientific research has driven the standards of
animal husbandry and behavior and has produced visible results. Animals in zoos
provide valuable data concerning the behavior of both captive and wild-living
populations (Barbosa, 2009; Watters, Margulis, & Atsalis, 2009). Since the 1980s,
much of the research conducted in zoos has involved the psychological health
of captive animals. Monitoring the animals' daily behaviors plays a significant
role in the assessment of animal well-being. The psychological health of captive
animals can have drastic consequences on its physiological health. To address the
psychological, social, nutritional, and physical health of the animal collection and
the public's view of zoos, modern zoos aim to display animals in what are deemed
naturalistic environments. Traditional cages are no longer considered healthy for the
animals and the visitor.

Zoological institutions have concluded that *in situ* behavior may not mimic
the behaviors normally seen in the organism's natural environment. For example,
zoological institutions have found that breeding success is not a determinant
of an animal's mental health. Some domestic animals breed readily in captive
environments and in close proximity to humans. This might explain why these
animals were selected for domestication. Domestic dogs will breed even in cramped
and unsanitary conditions, such as breeding farms and puppy mills. A number
of wild animals living in captivity also follow this breeding pattern, including
certain species of monkeys, anteaters, and birds. The previously named organisms
have been documented reproducing in small cages and under stressful stimuli.
Conversely, some animals such as the giant panda, mountain gorilla, and rhinoceros

have difficulty breeding in captivity and require assistance from zoo personnel. The welfare of the organism is a priority (Wickins-Drazilova, 2006), but breeding is paramount from a conservation standpoint. In fact, the *in situ* research at zoological institutions has made great advances in animal husbandry. As a result of the research conducted, zoos have reduced the need to capture wild animals to maintain their collections and are striving to release animals back into their natural environment (AZA, 2011d). The genetic exchange programs zoos have developed help maintain the genetic diversity of the collection (Bertram, 2004; Bostock, 1993; Watters et al., 2009; Wickins-Drazilova, 2006).

Even though zoological institutions have established that there is a relationship between long life and overall health (Bostock, 1993), the animals that live in zoos, on average, enjoy a longer lifespan than most captive animals. This is a result of being kept in a controlled environment with access to routine expert veterinary care. Zoo animals receive superior medical care and may play a key role in public health. Exotic animals are highly susceptible to foreign pathogens. Therefore, they are monitored daily through observations and blood tests. The results of the test can alert the animal management team and the public health organization of possible epidemics. For example, in 1999, wild crows began dying at an alarming rate, and epidemiologists were scrambling to find the cause, and exotic birds began to die at the Bronx Zoo. When the zoo-owned birds began to die, the zoo's veterinarian rushed to uncover the underlying cause in fear that other collections might become infected. The zoo uncovered the link between mass avian deaths and human disease when they determined that the West Nile virus was responsible for the birds' deaths. Even though zoos had played a vital role in the detection of the disease common in Africa and the Middle East, the results were not immediately released to the public. In 1999, zoos were not seen as a viable research institution and were considered disconnected from the mainstream public health (McNamara, 2007).

When animals are kept in captivity, they may exhibit undesirable and unnatural behaviors such as pacing, head swaying, and staring (Bostock, 1993; Wickins-Drazilova, 2006). The monitoring programs at zoos identify the circumstances in which stereotypic behaviors exist. As zoos have evolved so has their management of captive animals. They have been active in pursuing optimal care for the organisms and designing zoo exhibits. *In situ* research in zoological institutions provides a wealth of knowledge in captive breeding, behaviors, and effective environmental stimuli.

Animal husbandry, behavioral monitoring, and epidemiology are not the only scientific endeavors that take place within zoo collections. Contemporary research is concerning itself with climate change and global warming and the effects these have on captive and wild animals. As geologists have uncovered the geological patterns of long-term climate change, they have revealed periods of glaciation and extreme drought. Geologists believe glaciation and drought, in particular regions of the world, may have lead to megafaunal extinctions. Taking these geological patterns into consideration, zoological institutions are paying closer attention to the changes in the health of their animal collections and wild populations as they relate to environmental health. Although zoo animals live in controlled environments, they may offer some insight into parasitic infections and zoonotic diseases (Barbosa, 2009). As the climate warms, infections and diseases may become more prevalent. The Wildlife Conservation Society has found an increase of fly larva infecting baby birds in Argentina due to an increasingly muggy climate. The parasitic maggots burrow into the skin of nesting chicks and can kill the baby birds or cause abnormal growth (WCS, 2010).

The science of global warming and climatology is a relatively new research frontier for zoos. Because zoos have 150 years of meteorological, climatological, and geographical data (Barbosa, 2009), they are well positioned to be research leaders in climate change and its effects. We term this new science *bio-climatogeography*. Bio-climatogeography uses the meteorological data gathered by zoos and their geographical locations to determine how ecosystems, plants, and animals will react to global climate change and climatic patterns. Research into understanding the role zoos play in the critical issues of conservation is only beginning.

Culture and Society

Zoos are cultural institutions (Wharton, 2011). Throughout the historical development of zoos, they have been a part of society and have matured into important institutions that reflect current cultural and societal changes (Ballantyne et al., 2007; Fraser & Wharton, 2007; Hoage & Deiss, 1996; Marino, Lilienfeld, Malamud, Nobis, & Broglio, 2010). Zoos' architectural and exhibit designs reflect the past and present cultural impacts and subsequently project cultural perspectives to zoo visitors (Tarlow, 2001).

Society has and will probably continue to view zoological collections as centers of recreation. However, zoos perceive themselves as providing society with an enriching connection to conservation, biology, and organisms. Although the mission statements of zoos emphasize conservation, education, and research, zoos appear to be devoted to providing an enjoyable experience in a fun atmosphere. Now, zoos' goals need to focus on devising a plan that utilizes enjoyable, entertaining experiences to encourage informal education as the zoo increases much-needed revenue. Zoos need to utilize their marketing strategies and present information regarding their scientific activities. The style in which they choose to publicize their scientific and conservation endeavors must be interesting and entertaining. For example, the North Carolina Zoological Park produces ZooFilez with a local television news station. ZooFilez provides viewers with an opportunity to learn about organisms and the zoo's conservation efforts. According to a survey of 270 high school students living in the same county as the North Carolina Zoological Park, 90 students were aware of ZooFilez and stated their knowledge of zoos was gleaned from ZooFilez (Patrick & Tunnicliffe, 2009). Zoos can fulfill their commitments to conservation and research as well as deliver quality conservation education.

programming idea

The Future

Globalization and other environmental factors have led to a not-so-promising future. The relationship between human impact and habitat sustainability may not be easily conveyed through a simple visit to the zoo, but zoos must continue to communicate their message. Zoos hold the key to shaping the future relationship between society and nature. Most zoos have a limited capacity and are not able to manage large areas of natural habitats. However, all zoos assume five roles as the executor of the relationship between society and nature.

First, zoos take on the role of the "model citizen" by conveying a conservation message. They advocate for a sensible, sustainable use of natural resources and promote less wasteful, green-building alternatives. Many zoological institutions are developing organizational plans that include the use of solar, wind, and thermal power in their daily operations. Additionally, they are growing food for the animals, composting, and using recycled materials in their exhibit design. Second, zoos are maintaining a viable and genetically diverse collection. Zoos are managed under the premise that wildlife conservation is of foremost importance. As zoological institutions have become more active in field studies (i.e., dallaszoo.com/conservation/cs3_current), their research findings are being applied to larger conservation efforts. Moreover, the conservation research that takes place *in situ* and *ex situ* is important in saving small fragmented wild populations. Third, zoos directly influence the attitudes and behaviors of the community in relation to the conservation of plants, animals, and habitats. Due to their urban locations within heavily populated cities, zoos have a unique geographic placement within the community. The urban location of zoos provides them with a unique opportunity to influence government policy. Fourth, the zoo is a conservation mentor. Through mentoring efforts, future generations of scientists and citizens will be more aware of the benefits of long-term conservation (Rabb, 2004). As conservation mentors, zoos must lead the public to become citizen conservationist. Fifth, zoos are a place for people to learn basic facts about organisms and their behavior.

This is a technological era in which electronics rule our daily lives, and their use has become second nature. Information is available at the click of a button, but nothing can replace the mental health that interacting within nature provides (Louv, 2006). Even though most organisms in zoos are exotics, they do represent organisms in a simulated natural setting. Viewing animals in a naturalistic, though simulated, setting provides a sensory response that two-dimensional representations cannot duplicate (Broad & Smith, 2004; Rabb, 2004). Although it is difficult to measure the impact of a single zoo visit on an individual's behavior, zoos frequently ask visitors to participate in surveys that gauge the efficacy of the biological conservation message. Zoo visitors, even if it is their first visit, have preexisting knowledge of zoos (Patrick, 2010, 2011). A zoo visit may be the only interaction that a person living in an urban setting has with wildlife. Children are out of touch with nature, and their knowledge of organisms and biological conservation may be based on media, books, and formal classroom education (Falk et al., 2007; Patrick & Tunnicliffe, 2011; Rabb & Saunders, 2005). The number of zoo visitors each year looks promising. Each year, AZA-accredited institutions tout approximately 140 million visits. However, this statistic does not take into account repeat visitors who have a deeply rooted curiosity of organisms and biological conservation. How can zoos encourage people who are not interested in research, conservation, education, or the use of organisms for entertainment to visit? How can teachers utilize zoos as places for student research and education? Zoos and educators should capitalize on the human need to experience the diversity of the natural world and the knowledge people have of zoos.

References

Acampora, R. (1998). Extinction by exhibition: Looking at and in the zoo. *Human Ecology Review, 5*(1), 1–4.

Acampora, R. (2005). Zoos and eyes: Contesting captivity and seeking successor practices. *Society and Animals, 13*(1), 69–88.

Association of Zoos & Aquariums (AZA). (2011a, January 12). *AZA 5-year strategic plan*. Retrieved from http://www.aza.org/StrategicPlan/

Association of Zoos & Aquariums (AZA). (2011b, January 12). *The accreditation standards and related policies*. Association of Zoos and Aquariums. Retrieved from http://www.aza.org/accreditation/

Association of Zoos & Aquariums (AZA). (2011c, February, 8). *Animal care and management*. Retrieved from http://www.aza.org/animal-care-and-management/

Association of Zoos & Aquariums (AZA). (2011d, February 10). *Health, husbandry, and welfare*. Retrieved from http://www.aza.org/health-husbandry-and-welfare/

Ballantyne, R., Packer, J., Hughes, K., & Dierking, L. (2007, July). Conservation learning in wildlife tourism settings: Lessons from research in zoos and aquariums. *Environmental Education Research, 13*(3), 367–383.

Barbosa, A. (2009). The role of zoos and aquariums in research into the effects of climate change on animal health. *International Zoo Yearbook, 43*, 131–135.

Benirschke, K. (1987). Why do research in zoological gardens? *Canadian Veterinary Journal, 28*, 162–164.

Bertram, B. (2004). Misconceptions about zoos. *Biologist, 51*(4), 199–206.

Bitgood, S. (1988). A *comparison of formal and informal learning* (Technical Report No. 88–10). Jacksonville, AL: Jacksonville State University: Center for Social Design.

Bostock, S. S. (1993). *Zoos and animal rights: The ethics of keeping animals*. London: Routledge.

Broad, S., & Smith, L. (2004, February). *Who educates the public about conservation issues? Examining the role of zoos and the media*. International Tourism and Media Conference, Melbourne, Australia.

Cameron, S., Lozier, J., Strange, J., Koch, J., Cordes, N., Solter, L., & Griswold, T. (2011). Patterns of widespread decline in North American bumble bees. *Proceedings of the National Academy of Sciences of the United States of America, 108*(2), 662–667.

Carr, B. (2011, February 19). *Conservation Education in (AZA) Zoos and Aquariums*. Retrieved from http://www.izea.net/education/conservationed_aza.htm#top

Chizar, D., Murphy, J., & Lliff, N. (1990). For Zoos. *Psychological Record, 40*, 3–13.

Clarke, A. (2009). The Frozen Ark Project: The role of zoos and aquariums in preserving the genetic material of threatened animals. *International Zoo Yearbook, 43*, 222–230.

Clayton, S., Fraser, J., & Saunders, C. D. (2009). Zoo experiences: Conversations, connections, and concern for animals. *Zoo Biology, 28*, 377–397.

Convention on Biological Diversity. (2010). *Global diversity outlook 3*. Retrieved from www.cbd. int/GBO3

Conway, W. (2003). The role of zoos in the 21st century. *International Zoo Yearbook, 38*, 7–13.

Croke, V. (1997). *The modern ark. The story of zoos: Past, present and future*. New York: Scribner.

Dick, G. (2010, September). *Evolution of zoos*. Paper presented at the Association of Zoos and Aquariums Conference, Houston, TX.

Durrell Wildlife Conservation Trust. (2011, July 11). *Toadwatch*. Retrieved from http://www. durrell.org/In-the-field/Campaigns/Toadwatch/

Ellis, R. (2010, September). *Elephant barn open houses*. Paper presented at the Association of Zoos and Aquariums conference, Houston, TX.

European Association of Zoos and Aquariums (EAZA). (2011a, October 11). *EAZA collection planning*. European Association of Zoos and Aquariums. Retrieved from http://www.eaza.net/ activities/cp/Pages/Collection%20Planning.aspx

European Association of Zoos and Aquariums (EAZA). (2011b, October 11). *EAZA collection planning*. European Association of Zoos and Aquariums. Retrieved from http://www.eaza.net/ activities/Pages/Nutrition.aspx

European Association of Zoos and Aquariums (EAZA). (2011c, October 11). *EAZA collection planning*. European Association of Zoos and Aquariums. Retrieved from http://www.eaza.net/ activities/Pages/Research.aspx

Falk, J. H., & Dierking, L. D. (2000). *Learning from museums: Visitor experience and the making of meaning*. Walnut Creek, CA: Alta Mira Press.

Falk, J. H., Reinhard, E. M., Vernon, C. L., Bronnenkant, K., Heimlich, J. E., & Deans, N. L. (2007). *Why zoos and aquariums matter: Assessing the impact of a visit to a zoo or aquarium*. Silver Spring, MD: Association of Zoos and Aquariums.

Fernandez, E., & Timberlake, W. (2008). Mutual benefits of research collaborations between zoos and academic institutions. *Zoo Biology, 27*(6), 470–487.

Fonseca, C. R. (2009). The silent mass extinction of insect herbivores in biodiversity hotspots. *Conservation Biology, 23*(6), 1507–1515.

Fraser, J., & Wharton, D. (2007, January). The future of zoos. *Curator, 50*(1), 41–54.

Frost, W. (2010). *Zoos and tourism: Conservation, education, entertainment?* Bristol, UK: Channel View Publications.

Garner, J. (2010, September). *Elk Bugling, photo and keeper tours*. Paper presented at the Association of Zoos and Aquariums conference, Houston, TX.

Hancocks, D. (2001). *A different nature: The paradoxical world of zoos and their uncertain future*. Los Angeles, CA: University of California Press.

Hanna, J. (1996). Ambassadors of the wild. In N. Richardson (Ed.), *Keepers of the kingdom: The new American zoo*. Charlottesville, VA: Thomasson-Grant & Locke.

Hoage, R., & Deiss, W. A. (Eds.). (1996). *New animals: From Menagerie to Zoological Park in the Nineteenth Century*. London: The Johns Hopkins University Press.

Holst, A. (2010, September). *Zip lines and rope courses*. Paper presented at the Association of Zoos and Aquariums conference, Houston, TX.

Hutchins, M. (2003). Zoo and aquarium animal management and conservation: current trends and future challenges. *International Zoo Yearbook, 38*, 14–28.

Hutchins, M., & Thompson, S. (2008). Zoo and aquarium research: Priority setting for the coming decades. *Zoo Biology, 27*(6), 488–497.

International Zoo Educators Association (IZEA). (2011, February 18). *Conservation education theory and practice*. Retrieved from http://www.izea.net/education/conservationed.htm

Israel, M. (2011). How to position and package special tours and programs for revenue maximization. *How to Generate Revenue Through Your Web Site*. Association of Zoos and Aquariums, Atlanta, GA, 15 September 2011, Conference Presentation.

IUDZG/CBSG (IUCN/SSC). (1993). Executive Summary, The World Zoo Conservation Strategy; The Role of the Zoos and Aquaria of the World in Global Conservation.

Jordan, W., & Ormrod, S. (1978). *The Last Great Wild Beast Show: A Discussion on the Failure of British Animal Collections*. England: Constable.

Karkaria, D., & Karkaria, H. (1998). Zoorassic Park: A brief history of zoo interpretation. *Zoos' Print, 14*(1), 4–10.

Kellert, S., & Dunlap, J. (1989). *Informal learning at the zoo: A study of attitude and knowledge impacts*. Philadelphia, PA: Zoological Society of Philadelphia.

Koebner, L. (1994). *Zoo book: The evolution of wildlife conservation centers*. New York: Tom Doherty Associates.

Kolbert, C. (1995). What are we trying to teach? *Journal of the International Association of Zoo Educators, 32*, 6–9.

Leader-Williams, N., Balmford, A., Linkie, M., Mace, G. M., Smith, R. J., Stevenson, M., et al. (2007). Beyond the Ark: Conservation biologists' views of the achievements of zoos in conservation. In A. Zimmermann, M. Hatchwell, L. A. Dickie, & C. West (Eds.), *In Zoos in the 21st century: Catalysts for conservation?* (pp. 236–254). Cambridge, UK: Cambridge University Press.

Lessow, D. (1990). *Visitor perceptions of natural habitat zoo exhibits*. Unpublished doctoral dissertation, Indiana University, Bloomington, IN.

Linington, S. H. (2000). The Millennium Seed Bank Project. In B. S. Rushton, P. Hackney, & C. R. Tyne (Eds.), *Biological collections and biodiversity. Linnean Society occasional papers* (vol. 3, pp. 358–373). Otley: Westbury Publishing/London: Linnean Society.

Louv, R. (2006). *Last child in the woods: Saving our children from nature deficit disorder*. Chapel Hill, NC: Algonquin Books.

Ludwig, E. G. (1981). Study of Buffalo Zoo. In M. Fox (Ed.), *International journal for the study of animal problems*. Washington, DC: Institute for the Study of Animal Problems.

Marino, L., Lilienfeld, S., Malamud, R., Nobis, N., & Broglio, R. (2010). Do zoos and aquariums promote attitude change in visitors? A critical evaluation of the American zoo and aquarium study. *Society and Animals, 18*, 126–138.

Martin, S. (2000). *The value of shows*. Presented at the International Association of Avian Trainers and Educators National Conference, Memphis, TN.

Mazur, N., & Clark, T. (2001). Zoos and conservation: Policy making and organizational challenges. *Bulletin Series Yale School of Forestry and Environmental Studies, 105*, 185–201.

McNamara, T. (2007). The role of zoos in biosurveillance. *International Zoo Yearbook, 41*, 12–15.

Meretsky, V., Snyder, N., Beissinger, S., Clendenen, D., & Wiley, J. (2000). Demography of the California Condor: Implications for reestablishment. *Conservation Biology, 14*(4), 957–967.

Nichols, M. (1996). *Keepers of the kingdom: The new American zoo*. Richmond, VA: Thomasson-Grant and Lickle.

Ogden, J., & Heimlich, J. E. (2009). Why focus on zoo and aquarium education? *Zoo Biology, 28*, 357–360.

Owen, K., Murphy, D., & Parsons, C. (2009). ZATPAC: A model consortium evaluates teen programs. *Zoo Biology, 28*(5), 429–446.

Patrick, P. (2010, September). *What middle school students know about zoos*. Poster presented at the annual meeting of the Association of Zoos and Aquariums, Houston, TX.

Patrick, P. (2011, April). *Zoo acuity model: Middle level students' knowledge of zoos*. Paper presented at the National Association for Research in Science Teaching, Orlando, FL.

Patrick, P., Matthews, C., Ayers, D., & Tunnicliffe, S. (2007). Conservation and education: Prominent themes in mission zoo mission statements? *Journal of Environmental Education, 38*(3), 53–60.

Patrick, P., Matthews, C., Tunnicliffe, S., & Ayers, D. (2007). Mission statements of AZA accredited zoos: Do they say what we think they say? *International Zoo News, 54*(2), 90–98.

Patrick, P., & Tunnicliffe, S. (2009, November). *Zoo acuity model: Students' knowledge of the role of zoos in conservation*. Paper presented at the National Association of Biology Teachers Conference, Denver, CO.

Patrick, P., & Tunnicliffe, S. (2011). What plants and animals do early childhood and primary students' name? Where do they see them? *Journal of Science Education and Technology.* Invited article and Special Issue: Early Childhood and Nursery School Education. Available at: http://www.springerlink.com/content/e27121057mqr8542/

Price, E., Vining, J., & Saunders, C. (2009). Intrinsic and extrinsic rewards in a nonformal environmental education program. *Zoo Biology, 28*(5), 361–376.

Puan, C., & Zakaria, M. (2007). Perception of visitors towards the role of zoos: A Malaysian perspective. *International Zoo Yearbook, 41,* 226–232.

Rabb, G. B. (2004). The evolution of zoos from menageries to centers of conservation and caring. *Curator, 47*(3), 237–246.

Rabb, G., & Saunders, C. (2005). *The future of zoos and aquariums: Conservation and caring., 39*(1), 1–26.

Reed, T. (1973). *American Zoo and Aquarium Association Records, 1970–1994.* Archives, Manuscripts, Photographs Catalog. Washington, DC: Smithsonian Institution Research Information System.

Rogers, A. D., & Laffoley, D.d'A. (2011). *International Earth system expert workshop on ocean stresses and impacts* (Summary report). International Programme on the State of the Ocean, Oxford.

Smithsonian National Zoological Park (SNZP). (2011, January 14). *Recovery of the desert antelope.* Retrieved from http://nationalzoo.si.edu/SCBI/ReproductiveScience/AntelopesCervids/default.cfm

Sommer, R. (1972). What do we learn at the zoo? *Natural History, 81*(7), 29.

Sweet, C. (2006, January 11). *Mexican wolves, wild once again.* Retrieved from http://nationalzoo.si.edu/Publications/ZooGoer/2006/1/mexican_wolves.cfm

Tarlow, S. (2001). Decoding ethics. *Public Archaeology, 1,* 245–259.

Tribe, A. (2004). Zoo Tourism. In K. Higgenbottom (Ed.), *Wildlife tourism: Impacts, management and planning* (pp. 35–36). Melbourne: Common Ground.

Tunnicliffe, S. (1995). *Talking about animals: Studies of young children visiting zoos, a museum and a farm.* Unpublished doctoral dissertation, King's College, London.

Tunnicliffe, S. (1996). Conversations with primary school parties visiting animal specimens in a museum and zoo. *Journal of Biological Education, 30*(2), 130–141.

Turley, S. K. (1999). Conservation and tourism in the traditional UK zoo. *The Journal of Tourism Studies, 10*(2), 2–13.

Visscher, N., Snider, R., & Stoep, V. (2009). Comparative analysis of knowledge gain between interpretive and fact-only presentations at an animal training session: An exploratory study. *Zoo Biology, 28*(5), 488–495.

Wagoner, K., Chessler, M., York, P., & Raynor, J. (2009). Development and implementation of an evaluation strategy for measuring conservation outcomes. *Zoo Biology, 28*(5), 473–487.

Wagoner, B., & Jensen, E. (2010). Science learning at the zoo: Evaluating children's developing understanding of animals and their habitats. *Psychology and Society, 3*(1), 65–76.

Watters, J. V., Margulis, S. W., & Atsalis, S. (2009). Behavioral monitoring in zoos and aquariums: At tool for guiding husbandry and directing research. *Zoo Biology, 28,* 35–48.

Wharton, D. (2011). *Poetry and the Wild: The Language of Conservation Project.* Association of Zoos and Aquariums. Atlanta, GA. 15 September 2011. Conference Presentation.

Wickins-Drazilova, D. (2006). Zoo animal welfare. *Journal of Agricultural and Environmental Ethics, 19,* 27–36.

Wildlife Conservation Society (WCS). (2010, March 12). *Climate change: Prime time for parasites.* Retrieved from http://www.wcs.org/new-and-noteworthy/prine-time-for-parasites.aspx

World Association of Zoos and Aquariums (WAZA). (2011, February 20). *Environmental education.* Retrieved from http://www.waza.org/en/site/conservation/environmental-education

Chapter 4
Visitors' Knowledge of Zoos

For much of their history, zoos have affirmed only an imperial mastery over nature. When operated with intelligence and compassion, however, zoos can be a most effective conservation tools. How can they best achieve this? By becoming conservation centers; by placing nature preservation at the center of all their efforts; by reaching for the highest standards in all their projects and activities; by seeking to awake, enthrall, and educate; by articulating the wonderful benefits of conserving biological diversity; and by making strategic alliances with other cultural and natural history institutions. Zoos have the marvelous potential to develop a concerned, aware energized, enthusiastic, caring, and sympathetic citizenry. Zoos can encourage gentleness toward all other animals and compassion for the well-being of wild places....To help save all wildlife, to work toward a healthier planet, and to encourage a more sensitive populace are the goals for the new zoos (Hancocks, 2001, p. 252).

As purveyors of a critical conservation message, zoos have been thrust into the position of biological conservation educator. Even though research in how and what people learn in zoos has become a focus in the past 20 years (Askue, Heimlich, Yu, Wang, & Lakly, 2009; Boyle, 2005; Gwynne, 2004; Marino, Lilienfeld, Malamud, Nobis, & Broglio, 2010; Stevens, Sams, & Ogden, 2004), research needs to evaluate the basic biology and conservation knowledge visitors bring to the zoo. If zoo education programs pose a better understanding of what children know and think about the roles and purposes of zoological institutions, then zoo educators will have a construct upon which to scaffold educational material. The public's knowledge of zoos is addressed in this chapter by (1) identifying the zoo visitor and their reasons for visiting the zoo, (2) determining the visitor's perceptions of nature, (3) describing the importance of mental models, and (4) defining a model of the divergent and nonscientific knowledge and understandings people have of zoos. Providing a model of the knowledge or understandings people hold regarding zoos is an important tool that may be used when developing educational interactions.

Does it matter if people know why zoos exist and their role in conservation? Yes. If people are to support zoological institutions, then people must be aware of why zoos are important. Zoos may be seen as a place that consumes money instead of as an educational resource. Modern zoos rely on the financial support of the paying public. In fact, some zoological institutions rely on gate fees as their main

P.G. Patrick and S.D. Tunnicliffe, *Zoo Talk*, DOI 10.1007/978-94-007-4863-7_4,
© Springer Science+Business Media Dordrecht 2013

source of funding. This is most certainly true of zoos in the United Kingdom, which receive little or no government funding. Without funding, zoological institutions cannot perform their scientific and educational functions. Funding research into the visitor–institution relationship provides valuable information in exhibit design and aids zoos in becoming financially self-sufficient (Davey, 2006; Turley, 1999, 2001).

The Zoo Visitor and Their Reasons for Visiting the Zoo

Determining the definition of an average zoo visitor remains difficult and subjective because there is a lack of standard sampling designs that take into consideration the demographics of zoological institutions in various regions. First, early visitor studies focused mainly on the number of visitors that passed through the turnstiles. Second, there is no generalizable standard zoo design that is consistent across institutions. In the 1970s, visitor studies began an evolution. The changing foci of visitor studies coincided with the advancement of zoos as conservation organizations. While zoos were moving away from displaying taxonomic collections (i.e., primate houses), they were shifting their interests toward wildlife conservation, endangered species, and environmental issues. Research into who visits a zoo and why became more prevalent.

Even though there is no definite "who" when defining the zoo visitor, there are some consistencies in behavior and age of the visitors. Staying time at specific exhibits, attention to informational signage, and visitor movement and emotional response have been found to be similar for the majority of zoo visitors (Davey, 2006; Morgan & Hodgkinson, 1999). Research has shown that children comprise a large percentage of zoo visitors. Around 35% of visitors to wildlife attractions are children under the age of 15 (Cain & Meritt, 2007; Clayton, Fraser, & Saunders, 2009; Turley, 2001; Wineman, Piper, & Maple, 1996). Adult visitors frequently cite children as crucial in the decision to visit a zoo, and zoos are described by adults as child-oriented (Clayton et al., 2009; Turley, 2001). Children influence the repeat visits of family groups, the adult visitor's exhibit interest, and staying time at the exhibit. Moreover, visitors have stated that the "presence of children was significantly related to the importance to having a pleasurable day out" (Turley, p. 11).

Children, who make up the largest zoo audience, are most interested in enclosures that contain animals of large body size, exhibit a large number of animals, and display animals' social behaviors (Ward, Mosberger, Kistler, & Fischer, 1998). Additionally, seeing "pretty animals," is one of the reasons cited by people for visiting a zoo (Kellert, 1980). Zoo visitors are drawn to certain zoo animals, such as the giant panda, because of the animals' anthropomorphic features (Morris, 1961; Surinova, 1971). As zoos tout their educational programs, rich research interests, and scientific and conservation beneficence, their exhibit design and educational programs appear to be driven by *pretty animals*. However, some institutions have specific programs geared toward children's cognitive development and/or biological conservation awareness. Zoos, such as the Dallas Zoo and the Denver Zoo, are striving to influence the next generation of scientists and conservation savvy citizens. The Dallas Zoo offers Nature Tykes for 3–4-year-olds and A–Zoo Preschool Safari for children ages 4–5 years. Nature Tykes provides a certified teacher who works with children to develop their science thinking and problem-solving skills. A–Zoo Preschool Safari uses the alphabet and literacy to explore animals (Dallas Zoo, 2011). Children at the Denver Zoo may choose to participate in Tot Trekkers (ages 2–3 years), Career Discovery Day (ages 10–18 years), and Zoo Crew (ages 13+). Tot Trekkers includes play-based activities for children and parents to promote learning about animals. Career Discovery Day provides opportunities for participants to interact with zoo professionals and gather information concerning zoo careers. Zoo Crew is a volunteer program for teenagers during which the teenagers shadow a zoo keeper.

The public's reasons for visiting zoos hold important implications for zoo management. The first step in identifying why people visit is understanding visitor identity. Falk et al. (2007) and Falk (2009) determined that there are five visitor identities: (1) Facilitators represent the largest percentage of visitors. They desire a more social experience and visit with another person. (2) Explorers associate their experiences with the concept of learning. They believe their experiences become educational through observing and interpreting animals and their behavior. (3) Experience seekers perceive zoos as a tourist attraction or a community activity. They have the lowest expectations for their visits. (4) The professional/hobbyist has the highest amount of prior knowledge of all zoo visitors. They are more familiar with the zoo and are more likely to participate in specialized programs and volunteer. (5) Spiritual pilgrims relate their visit to a moment of self/spiritual reflection. However, studies show that "science talkers" is not one of the five identities. Discourse that occurs at the zoo does not include science terms. Visitors look at animals and label them by allocating them to categories with which they are already familiar. Visitors view the animals in isolation and do not compare the animals with each other, do not notice the animals' salient features and behaviors, and do not discuss the animals' diet or natural habitat (Tunnicliffe, 1996a, 1996b, 1996c, 1997). The following conversation took place in the Arabian Oryx exhibit. The Mother recognizes the goat features of the Oryx and the mother names the animal as a goat:

Arabian Oryx

Mother: Oh look! It's a goat!
Grandmother: Oh, yes. Look! A goat. [She is saying this to her granddaughter.]
Mother: It must be a goat because it has hair, hooves, and horns.

Falk et al.'s (2007) and Falk's (2009) results reinforce the findings of Turley (2001). Turley asked people why they visit a zoo. The study establishes six main reasons, which influence zoo visits. Recreation was stated more often than any other reason for deciding to visit a zoo. Visitors want to have a "pleasurable day out" and "experience animals close-up," which can be directly related to Falk et al.'s facilitator identity. Coinciding with Falk et al.'s explorer identity, the second most-named reason was education and experiencing wild animals. Some research claims that there is actually no education taking place in zoos (Acampora, 2005; Bertram, 2004), but "it would appear that the role of education has the greatest appeal among the core visiting market" (Turley, p.10). Additionally, Falk et al. and Turley found that discovering, understanding, and supporting the conservation activities of the zoo were not the number one priority of visitors (i.e., professional/hobbyist identity).

The Visitor's Perceptions of Nature

According to E.O. Wilson's biophilia hypothesis, humans have an innate desire to catalog, understand, and spend time with other life-forms (Kellert & Wilson, 1993). The biophilia hypothesis states that all people are characteristically drawn to nature. They need to have an affiliation with nature in order to succeed and obtain an optimal level of self-value. Wild animals inspire our innate caring about species and nature because of their dependence and beauty and because we relate to them as sensing, perceiving creatures. People do come in contact with animals as pets or as urban wildlife (Tunncliffe & Reiss, 1999), such as squirrels, trees, grasses, deer, opossums, raccoons, birds, and ants, and have a grasp of these organisms' ecological relationships (Boulter, Tunnicliffe, & Reiss, 2003). However, owning a pet only provides people with a limited knowledge of organisms (Ascione, 1992; Kellert, 1980), and industrialization and urbanization are reducing human's direct interactions with nonurban nature (Balmford, 1999; Kellert & Wilson, 1993). More-over, children (8–9 years old) in urban Chicago were asked about their perceptions of their urban environment. The children's perceptions of the urban environment were positive, but they perceived nature outside the urban environment as harmful or dangerous (Simmons, 1994). Despite this fear of the nonurban world, some research shows that students believe the nonurban natural world is important. For example, students interviewed in an impoverished community in Houston, Texas, said that animals (84%), plants (87%), and parks/open spaces (70%) were an important part of their lives. Results also showed that it would matter to these children if polluting the bayou harmed birds (94%), water (91%), insects (77%), and the view (93%) (Kahn & Friedman, 1995). As the world becomes more urbanized, our personal experiences with animals will become more isolated. Lacking first-hand experiences with nonurban wildlife, much of the urban-American population has come to depend upon wildlife organizations, publications, and television programs for wildlife education (Wilkinson, 1997).

Research links the loss of knowledge about the nonurban natural world to growing isolation from it (Kellert & Wilson, 1993; Nabhan & Trimble, 1994). Ewert and McAvoy (1994), Louv (2006), and Shepard (1982) believe that many environmental, social, psychological, and educational problems are rooted in the lack of public knowledge about natural resources. As early as 1969 (Levinson), psychologists believed that as humans separated themselves farther from nonurban nature, the healing forces of nature and the animal kingdom were being left behind. For this reason, conservationists need to reestablish children's links with nonurban nature if they are to bridge the gap between children and their desire to conserve (Balmford, Clegg, Coulson, & Taylor, 2002). To quantify the knowledge children have of nature and the effects of human-made creatures on their innate interest in diversity, students in the United Kingdom were shown pictures of organisms and Pokémon flashcards and asked their names. At age 8, students could identify nearly 80% of a sample drawn from 150 synthetic Pokémon species. However, when asked to name the real organisms, they named less than 50%. The creators of Pokémon are doing a better job than conservationists at inspiring interest in students (Balmford et al., 2002).

People who belong to wildlife and/or environmental organizations, watch animal-related television programs, read books, take nature walks, and visit places of informal learning have a higher base of nature-related knowledge than those who do not (Boulter et al., 2003; Eagles & Muffitt, 1990; Falk & Dierking, 2010; LaHart, 1978; Tunnicliffe & Reiss, 2000; Westervelt & Llewellyn, 1985). School is not listed here as a main source of nature learning because research shows that school plays a very small role in pupils' recollections of their sources of learning (Boulter et al., 2003; Falk & Dierking, 2010; Patrick & Tunnicliffe, 2011; Tunnicliffe & Reiss, 2000). Therefore, out of classroom, informal interactions and activities are important in the development of the knowledge of organisms and nature. If people need informal, natural interactions to learn about organisms and appreciate nature, then these interactions are an important part of understanding the importance of conservation and developing a regard for preservation.

People have inaccurate perceptions regarding organisms (Bitgood, Formwalt, Zimmerman, & Patterson, 1993; Patrick & Tunnicliffe, 2011; Tunnicliffe, 2000; Tunnicliffe & Reiss, 2000) and zoos' roles in conservation (Patrick, 2010; Patrick & Tunnicliffe, 2009). Therefore, before zoos can correct the public's attitudes and perceptions and change the public's behavior, zoos need to better understand the visitor and the visitor's mental models of zoos.

The Importance of Mental Models

Mental models provide a framework for interpreting and predicting various phenomena (Vosniadou, 1998) and for determining appropriate responses to new situations, as well as for guiding the perceptions, decisions, and behaviors of an individual (Kearney & Kaplan, 1997). Kearney and Kaplan (1997) describe mental models as hypothesized knowledge structures symbolizing the misconceptions, assumptions, beliefs, and perceived facts a person holds about the world. Our mental models become what we abstract from our experiences through personal construct and store in memory as an example of something or some situation (Garnham, 1997).

A person's perceptions and interactions influence the way they evaluate the external world (LaHart, 1978). Therefore, the way a person interacts within the world, interprets new information, and makes meaning of the new information is a process of learning. As educators gain knowledge of what students bring to a situation, educators gain insight on how learning environments can be more effectively designed. Ausubel (1968) emphasized the importance of prior ideas and knowledge in the construction of new knowledge. Not just as a structure to be modified but as strongly held, ingrained ways of seeing the world. The interaction of prior ideas with new ideas through a social process is described by Smith (1991) as conceptual change theory. The interactions of the mental and scientific consensus models of phenomena have moved researchers toward an appreciation of learning as a social process. Education as a social process allows teachers to link personal experiences, classroom activities, and ideas in a curriculum structured to support learning.

Visitors orient themselves to a place by applying their prior knowledge, attitudes, and expectations. Therefore, the more educators know about students' prior knowledge, including students' mental models, the better-prepared educators will be able to aid students in adjusting their prior knowledge to reflect correct mental models (Coburn, 1995). Therefore, if zoos are going to change visitors and nonvisitors conservation-related knowledge and behavior, then zoos must better understand the mental models of the individual visitors.

When researchers, conservation biologists, or educators ask a person about their understandings, the person will respond by describing their own personal mental model (Bakhurst & Shanker, 2001). The description may be words, drawings, physically constructed objects, mathematical symbols, or even gestures and are usually considered to be the person's expressed models (Buckley, Boulter, & Gilbert, 1997). Therefore, the expressed model of a mental model is a person's beliefs about a phenomenon that are placed in a public domain. However, an individual's complete, comprehensive mental model is inaccessible to all but the owner. The only way an individual gains some understanding of another person's mental model is by accessing one or more of their expressed models.

Even though mental models are considered an individual's perceptions, thoughts, beliefs, and understandings, these individual understandings can be grouped collectively to define the mental model of a group (Durkheim, 1938). Therefore, this book looks at the Zoo Voice of visitors to determine the group mental model of zoo visitors. The group mental model may not represent every single individual mental model, but "the same type of processes that occur for individuals are conceptually involved in the information processing by the group" (Hinsz, 1990, p. 12). Because the concept of "zoo" is a socially shared concept that is composed of individual socially shared belief structures (e.g., Damon, 1991; Resnick, 1991), then understanding these individual beliefs as a group will strengthen our understanding of the Visitor Voice as a collective.

To understand students' mental models of places of informal learning, we need to define their mental models and identify the influences on their mental models. Mental models are impacted by external stimuli and the way those stimuli are

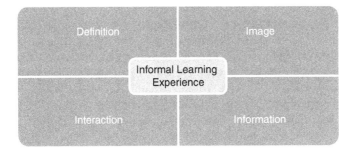

Fig. 4.1 Informal learning model

assimilated into the person's current understandings. The mental models people hold of places of informal learning are defined by the informal learning model (Fig. 4.1): (1) definition, (2) image, (3) interaction, and (4) information. The student's knowledge is the mental model that is molded by the informal learning experience. The definition of what happens during a visit is demarcated by accreditation organizations such as AZA, EAZA, IZE, and WAZA; the historical perspective (see Chap. 2); and the zoo's mission statement (see Chap. 3). The visual image is depicted through the zoo's exhibits, staff, and outreach (i.e., television, websites, advertisement, visits to schools, etc.). The interactions that impact the mental models of students are the conversations that take place before, during, and after the visit (i.e., teachers, parents, friends, staff). Additionally, students' mental models are shaped by the information the zoo provides and information that is gleaned from other nonzoo sources such as the Internet and television. For zoos to make an impact in learning during a zoo visit, an understanding of the mental models students hold of zoos is important. As informal learning occurs, the mental model is adjusted or changed to accommodate the new information.

An individual's affective domain, cognitive domain, existing knowledge, and culture influence their existing attitudes to a situation. Changing the mental models people hold is a long and difficult process that occurs incrementally. A talk provided at an exhibit or an exhibit that tells a narrative that conflicts with a person's existing mental constructs will not exact change. Yet, zoos rely on talks, shows, and exhibits to explain their biology conservation mission and inspire change. Instead, a gradual partnership of understanding needs to be developed between the conservationists and the visitor's zoo knowledge and conservation concerns. Therefore, understanding how learning theory and practice play a role in developing new knowledge and identifying conservation psychology are important in effecting change in visitors' beliefs.

The initial role of the researcher or teacher is to assess students' prior knowledge. By linking new concepts to previous ones or by causing students' dissatisfaction with their current schema, the process of conceptual change begins. If a constructivist learning theory is to be followed in zoo education, then uncovering students' current knowledge structures is the first step. In developing complex ideas, such as biological conservation and the zoo's role in conservation, students' mental models are important. Once the gaps in students' mental models of biological conservation, zoos, and the role of zoos in biological conservation are determined, the information needed to fill the conceptual gap can be developed (Young, 2008). Not only is it important to find the gaps within the mental models, but it is also important to provide "…novice learners [visitors in this book] with supportive information before practice can contribute substantially to the progression of a learner's mental model toward an expert-like mental model" (Darabi, Nelson, & Seel, 2010, p. 115). If learning, in terms of constructing understanding, is to occur during a zoo visit, learners must be able to place the experience into a meaningful conceptual framework. Hence, the content of visitors' knowledge or mental models prior to the zoo visit is of great importance in developing educational opportunities. A clearer glimpse into the student's understanding of zoos and beliefs about how zoos function in relation to conservation can provide fundamental information about the knowledge and agenda of visitors.

The zoo education staff is ultimately responsible for creating the opportunities for learning that may arise from the experience of the visit (Kellert & Westervelt, 1983). However, visitors come to the zoo with an array of experiences and lifelong constructed knowledge. Often, zoos need to correct misinformation before new or desired learning can occur (Borun, Massey, & Lutter, 1992). Attractions themselves present experiences, and it is the nature of an experience to be determined and interpreted largely by the individual (Boud, Keough, & Walker, 1985). Therefore, describing each of the components of the informal learning mental model is important. We know how zoos define themselves through the (Association of Zoos & Aquariums (AZA), 2011) and their mission statements (Patrick, Matthews, Ayers, et al., 2007; Patrick, Mattews, Tunnicliffe, et al., 2007). Much research has been done to describe exhibits and their educational impact; explain how teachers, staff, and chaperones influence student learning during a zoo visit; and look at where students glean information. However, no one has defined the mental model students have of zoos. Once a clearer picture of the mental model is expounded, zoos and educators will have a clearer idea of how to design the individual elements (definition, image, interaction, information) that shape informal education and influence the mental model.

Understandings People Have of Zoos

Zoos have become more aggressive in their education programs and have moved away from strictly taxonomic and natural history themes toward ecological interpretation and conservation implications (Ballantyne, Packer, Hughes, & Dierking, 2007; Frost, 2010; Hunt, 1993; Ogden & Heimlich, 2009; Thurston, 1995). To promote their messages, nearly all zoos, including small zoos, in the USA, Europe, and Australia have education departments (Heimlich, 1996; Walker, 1991). However, having an education staff, a living animal collection, and attractive signage does not ensure that education is automatic (Croke, 1997; Walker, 1991). Zoos must be aware of how their conservation messages affect the visitors, visitors' prior knowledge and knowledge gained during the visit, the feelings that motivate or suppress visitors' conservation action, the visitor's reactions to the excitement of the visit, and the understandings visitors have of biological conservation.

Educators assume that their biological conservation education initiatives automatically bring an individual closer to nature, which, in turn, positively affects her/his environmental ethics (Bogner, 1999). The ultimate goal of education in zoos is to generate, develop, and raise knowledge, awareness, and concern about nature because people care about what they know (Bogner, 1999; Habban & Trimble, 1994; Kellert & Wilson, 1993). Education in zoos strives to help people realize that they share the world with other beings and develop children's awareness of nature and their confidence, empathy, and respect for others (Hancocks, 2001; Naherniak, 1995). Zoos' conservation education programs have the opportunity to fill the gap between education, awareness, and behavior and focus children's ideas concerning saving wild places and wildlife.

The primary reason cited for visiting a zoo is claimed to be for the educational benefit of children (Kellert, 1980). However, zoos have to be aware of their educational limitations. The knowledge scores of those who visit a zoo are not significantly different from those of nonvisitors (Kellert, 1980). Through good communication skills, zoos have the opportunity to educate the public on the needs of saving animal species from extinction and preventing habitat destruction (Hutchins & Conway, 1995). Understanding visitors' prior knowledge of animals and biological conservation has been an important aspect of educating zoo visitors and affecting their learning experiences (e.g., Falk et al., 2007; Moss, Esson, & Bazley, 2010; Ross & Gillespie, 2009; Smith, Curtis, & Van Dijk, 2010). A plea for a better understanding of the knowledge the public has of zoos was set forth by Paul Boyle at the 2005, AZA National Conference. Boyle stated

> We need to know our audience better . . . We need to do more research to understand our audience. We need to understand where our public is coming from. We need research which documents what people know and what they understand. We need research which gives a picture of how people understand conservation and the role of conservation organizations. What do they bring? What level is their understanding? How much information can we give them and not compromise their level of understanding? The only way to accomplish this is to understand our audiences. (Boyle, 2005)

Wineman et al. (1996) have shown that students are able to describe zoo exhibits in relation to their environmental concerns but did not determine if students actually thought of zoos as biological conservation organizations. Students were not asked if they viewed zoos as conservation organizations nor were they asked about the role of zoos in biological conservation. In fact, students reiterated their beliefs that zoos are a place for socializing.

Students, teachers, and the general public have a weak base of conservation-related knowledge, attitudes, and practices (Birney, 1991; Bowers, 1997; Price, Ashmore, & McGivern, 1994; Swanagan, 2000). There is a growing recognition that human understanding of the natural environment must be altered through biological conservation education. The manner in which zoos choose to approach the task of biological conservation education must be taken seriously. The education staff (Chizar, Murphy, & Lliff, 1990), the exhibits (Boud et al., 1985), the zoo experience, and the prior knowledge of the visitor (Chizar et al., 1990; Gwynne, 2004) are ultimately responsible for creating opportunities for learning that may arise from the zoo experience. The outcome of a zoo visit is situated in the affective domain of learning, the domain concerned with feelings, emotions, beliefs, and attitudes (Jenner, 2003). In some instances, attitudes toward the environment and animals are more negative after a visit to a zoo (Swanagan, 2000). In other cases, children who visit the zoo use positive emotional terms to describe their visit and the interactions they have with zoo personnel during their zoo visit (Gwynne, 2004; Stevens et al., 2004), implying that the presence of live animals and their caretakers influences the tenor of their responses. Not only is learning affected by expectations and social influences but by the physical environment as well (Birney, 1988). Many researchers stress the need for studies that document the impact of visits to zoos on biological conservation knowledge, awareness, and behavior (Churchman & Marcoulides, 1991; de White & Jacobson, 1994). A study conducted by the Chicago Zoological Society and the Lincoln Park Zoological Society (1993) concluded that zoo visitors do not think of zoos as conservation organizations.

The literature fails to examine visitors' knowledge of zoos and the roles of zoos in conservation. Therefore, Patrick (2010) and Patrick and Tunnicliffe (2009) asked 538 10–14-year-olds (Patrick) and 234 15–18-year-olds (Patrick & Tunnicliffe) to identify where they learned about zoos. Of the 772 students in the studies, 41% named the zoo as the source of information about zoos. Moreover, using questionnaires and concept mapping (i.e., Why do zoos exist? How are zoos involved in conservation?), the two studies described student's zoo knowledge. The studies revealed seven concepts students used to describe zoos. The concepts were the following, in descending order followed by the number of students who mentioned the concept: (1) *organisms* (771), (2) *people* (255), (3) *amenities* (510), (4) *descriptive terms* (582), (5) *habitats* (70), (6) *education* (353), and (7) *conservation* (341). The seven themes fell into three categories: observations (*organisms* and *descriptive terms*), interactions (*people* and *amenities*), or information (*education*, *conservation*, and *habitat*). Three hundred and forty-one of the students in both studies used the word conservation or a conservation-related term, that is, save animals. When students were asked, "Why do zoos exist?" 62 students mentioned conservation or a conservation-related term. The additional conservation-related words were used when the students answered the question, "How are zoos involved in conservation?"

The results have been used to devise the zoo knowledge model (Fig. 4.2). The zoo knowledge model is a visual of students' zoo knowledge. The observation category and the interaction category contain richer answers. Eighteen percent of the words or phrases used to describe zoos were related to the information category. These studies demonstrate that students do not fully understand the purpose of zoos and are not able to explain how zoos are involved in biological conservation. It is now imperative to utilize the observation and interaction categories to begin building

Fig. 4.2 Zoo knowledge
model

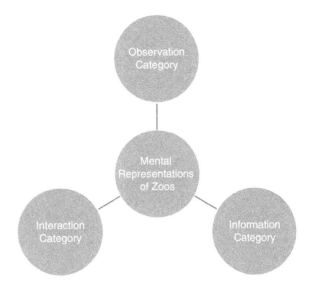

a stronger interrelationship between all three of the categories. Now, it is the responsibility of zoos to form a link between the understandings students have of zoos and saving habitats.

Even though zoos identify themselves as conservation organizations, which utilize education as a catalyst for ecological change, the participants in the studies described above do not view zoos as conservation organizations. Zoological programs have been developed to educate the public on matters involving conservation, animal care, and research, but most visitors still visit to see animals and be entertained. Research conducted by Clayton et al. (2009) further support these findings. "Visitors are receptive to the opportunities for learning afforded by the zoo, but only insofar as it fits within their goals for enjoyment in a social setting" (Clayton et al., p. 393). Visitors perceive zoos as recreational centers, where social interactions take place. Moreover, "zoos are perceived by their visitors to be valuable venues for social experiences" (Clayton et al., p. 393).

Identifying the public's knowledge of zoos is a key concept for properly marketing and enhancing the zoo experience. Currently, zoo visitors see zoos as entertainment venues. Informing the public about what zoos do and their role in biological conservation efforts is a daunting task. Zoo directors must develop a system which emphasizes zoos' goals of education, conservation, and research. Zoos are utilizing *edutainment* to goad visitors into learning about conservation. For example, increasing the visitor's time at the zoo through zip wire adventures is not going to increase the visitor's conservation awareness. In fact, some *edutainment* activities may result in visitors leaving the zoo with less knowledge of conservation. Research needs to be conducted which supports a relationship between these new *edutainment* approaches to teaching conservation and the knowledge gained. The ensuing chapters provide more information concerning the four components of the informal learning model.

References

Acampora, R. (2005). Zoos and eyes: Contesting captivity and seeking successor practices. *Society and Animals, 13*(1), 69–88.

Ascione, F. (1992). Enhancing children's attitudes about the humane treatment of animals: Generalized to human-directed empathy. *Anthrozoos, 5*(3), 176–191.

Askue, L., Heimlich, J., Yu, J., Wang, X., & Lakly, S. (2009). Measuring a professional conservation education training program for zoos and wildlife parks in China. *Zoo Biology, 28*(5), 447–461.

Association of Zoos & Aquariums (AZA). (2011, January 12). *AZA 5-year strategic plan.* Retrieved from http://www.aza.org/StrategicPlan/

Ausubel, D. P. (1968). *Educational psychology: A cognitive view.* New York: Holt, Rinehart & Winston.

Bakhurst, D., & Shanker, S. (2001). *Jerome Bruner: Language, culture and self.* New York: Sage Publications.

Ballantyne, R., Packer, J., Hughes, K., & Dierking, L. (2007, July). Conservation learning in wildlife tourism settings: Lessons from research in zoos and aquariums. *Environmental Education Research, 13*(3), 367–383.

Balmford, A. (1999). (Less and less) great expectations. *Oryx, 33*, 87–88.

Balmford, A., Clegg, L., Coulson, T., & Taylor, J. (2002). Why conservationists should heed Pokémon? *Science, 295*, 2367–2371.

Bertram, B. (2004). Misconceptions about zoos. *Biologist, 51*(4), 199–206.

Birney, B. (1988). Brookfield zoo's flying walk exhibit: Formative evaluation aids in the development of an interactive exhibit in an informal learning setting. *Environment & Behavior, 20*(4), 416–434.

Birney, B. (1991). *The impact of bird discovery point on visitors' attitudes toward bird conservation issues.* Brookfield, IL: Chicago Zoological Society.

Bitgood, S., Formwalt, D., Zimmerman, C., & Patterson, D. (1993). The Noah's Ark dilemma: zoo visitors' ratings of how much animals are worth saving. *Journal of the International Association of Zoo Educators, 27*, 41–43.

Bogner, F. (1999). Empirical evaluation of an educational conservation program introduced in Swiss secondary schools. *International Journal of Science Education, 21*(11), 1169–1185.

Borun, M., Massey, C., & Lutter, T. (1992). *Naive knowledge and the design of science museum exhibits.* Philadelphia: Franklin Institute Science Museum.

Boud, J., Keough, R., & Walker, D. (1985). *Understanding your visitors: Ten factors that influence visitor behavior.* Jacksonville, AL: Jacksonville State University.

Boulter, C., Tunnicliffe, S., & Reiss, M. (2003). *Probing children's understanding of the natural world.* Paper presented at the National Association for Research in Science Teaching Conference, Philadelphia, PA.

Bowers, C. (1997). *The culture of denial: Why the environmental movement needs strategy for reforming universities and public schools.* New York: Suny.

Boyle, P. (2005, April 14). *Association of zoos and aquariums conference.* Knoxville, TN. Speaker.

Buckley, B., Boulter, C., & Gilbert, J. (1997). Towards a typology of models for science Education. In Gilbert, J. (Ed.), *Exploring models and modelling in science and technology education* (pp. 90–105). Reading, MA: University of Reading New Bulmershe Papers.

Cain, L., & Meritt, D. (2007). The demand for zoos and aquariums. *Tourism Review International, 11*(3), 295–306.

Chicago Zoological Society and Lincoln Park Zoological Society. (1993). *Conservation related perceptions, attitudes, and behavior of adult visitors and nonvisitors to Brookfield Zoo and Lincoln Park Zoo.* Unpublished manuscript.

Chizar, D., Murphy, J., & Lliff, N. (1990). For zoos. *Psychological Record, 40*, 3–13.

Churchman, D., & Marcoulides, G. (1991). *Affective response to zoo exhibits.* Paper presented at the Association of Zoological Parks and Aquariums conference, Wheeling, WV.

Clayton, S., Fraser, J., & Saunders, C. D. (2009). Zoo experiences: Conversations, connections, and concern for animals. *Zoo Biology, 28*, 377–397.

Coburn, W. (1995). *Everyday thoughts about nature: An interpretive study of 16 ninth graders' conceptualizations of nature.* In Proceedings of Annual Meeting of the National Association for Research in Science Teaching, San Francisco.

Croke, V. (1997). *The modern ark: The story of zoos, past, present and future.* New York: Scribner.

Dallas Zoo. (2011, May 17). *Conservation education and science.* Retrieved from http://www. dallaszooed.com/kidcamps/preschool.html

Damon, W. (1991). Problems of direction in socially shared cognition. In L. B. Resnick, J. M. Levine, & S. D. Teasley (Eds.), *Perspectives on socially shared cognition* (pp. 384–397). Washington, DC: American Psychological Association.

Darabi, A., Nelson, D., & Seel, N. (2010). Learning and instruction in the digital age. In J. Spector, D. Ifenthaler, P. Isaias, & D. Sampson (Eds.), *The role of supportive information in the development and progression of mental models.* Dordrecht, The Netherlands: Springer.

Davey, G. (2006). An hourly variation in zoo visitor interest: Measurement and significance for animal welfare research. *Journal of Applied Animal Welfare Science, 9*(3), 249–256.

de White, G., & Jacobson, S. (1994). Evaluating conservation education programs at a South American zoo. *Journal of Environmental Education, 25*(4), 18–23.

Durkheim, E. (1938). *Rules of sociological method.* New York: Free Press.

Eagles, P., & Muffitt, S. (1990). An analysis of children's attitudes toward animals. *Journal of Environmental Education, 21*(3), 41–44.

Ewert, A., & McAvoy, L. (1994). Outdoor education research: Implications for social/educational and natural resource policy. In L. McAvoy, L. Stringer, & A. Ewert (Eds.), *Coalition for education in the outdoors second research symposium proceedings.* Cortland, NY: Coalition for Education in the Outdoors.

Falk, J. (2009). *Identity and the museum visitor experience.* Walnut Creek, CA: Left Coast Press.

Falk, J., & Dierking, L. (2010). The 95 percent solution: School is not where most Americans learn most of their science. *American Scientist, 98*, 486–493.

Falk, J. H., Reinhard, E. M., Vernon, C. L., Bronnenkant, K., Heimlich, J. E., & Deans, N. L. (2007). *Why zoos and aquariums matter: Assessing the impact of a visit to a zoo or aquarium.* Silver Spring, MD: Association of Zoos and Aquariums.

Frost, W. (2010). *Zoos and tourism: Conservation, education, entertainment?* Bristol, UK: Channel View Publications.

Garnham, A. (1997). Representing information in mental models. In M. Conway (Ed.), *Cognitive model of memory.* Cambridge, MA: MIT Press.

Gwynne, J. (2004). *Inspiration for conservation: Motivating audiences to care.* Presented in Proceedings Catalysts for Conservation: A Direction for Zoos in the 21st Century. London Zoo: Zoological Society of London.

Habhan, G. P., & Trimble, S. (1994). *The geography of childhood: Why children need wild places.* Boston: Beacon.

Hancocks, D. (2001). *A different nature: The paradoxical world of zoos and their uncertain future.* Los Angeles: University of California Press.

Heimlich, J. E. (1996). *Adult learning in nonformal institutions.* Columbus, OH: ERIC Clearinghouse on Adult, Career, and Vocational Education.

Hinsz, V. B. (1990). *A conceptual framework for a research program on groups as information processors.* A technical report submitted to the Logistics and Human Factors Division, AF Human Resources Division, AF Human Resources Laboratory, Wright-Patterson AFB, OH.

Hunt, G. (1993). Environmental education and zoos. *Journal of the International Association of Zoo Educators, 29*, 74–77.

Hutchins, M., & Conway, W. (1995). Beyond Noah's Ark; The evolving role of modern zoological parks and aquariums in field conservation. *International Zoo Yearbook, 34*, 117–130. London: The Zoological Society.

Jenner, N. (2003). *A communications strategy for Jersey Zoo within the context of the visiting public.* Report to the Durrell Wildlife Conservation Trust, Jersey Island, UK.

Kahn, P., & Friedman, B. (1995). Environmental views and values of children in an inner-city Black community. *Child Development, 66*, 1403–1417.

Kearney, A., & Kaplan, S. (1997). Toward a methodology for the measurement of knowledge structures of ordinary people: The conceptual content cognitive cap (3CM). *Environment and Behaviour, 29*(5), 579–589.

Kellert, S. R. (1980). *Phase II: Activities of the American public relating to animals*. United States Department of the Interior Fish and Wildlife Service.

Kellert, S., & Westervelt, M. (1983). *Children's attitudes, knowledge and behaviors toward animals: Phase V*. Washington, DC: U.S. Department of the Interior Fish and Wildlife Service.

Kellert, S., & Wilson, E. (1993). *The biophilia hypothesis*. Washington, DC: Island Press.

LaHart, D. E. (1978). *The Influence of knowledge on young people's perceptions about wildlife*. Unpublished doctoral dissertation, Florida State University, College of Education, Tallahassee, FL.

Levinson, B. (1969). *Pet-oriented child psychotherapy*. Springfield, IL: Charles C. Thomas.

Louv, R. (2006). *Last child in the woods: Saving our children from nature deficit disorder*. Chapel Hill, NC: Algonquin Books.

Marino, L., Lilienfeld, S., Malamud, R., Nobis, N., & Broglio, R. (2010). Do zoos and aquariums promote attitude change in visitors? A critical evaluation of the American zoo and aquarium study. *Society & Animals, 18*(2), 126–138.

Morgan, J. M., & Hodgkinson, M. (1999). The motivation and social orientation of visitors attending a contemporary zoological park. *Environment & Behavior, 31*, 227–239.

Morris, D. J. (1961). Automatic at London Zoo seal-feeding apparatus at London Zoo. *International Zoo Yearbook, 2*(1), 70.

Moss, A., Esson, M., & Bazley, S. (2010). Applied research and zoo education: The evolution and evaluation of a public talks program using unobtrusive video recording of visitor behavior. *Visitor Studies, 13*(1), 23–40.

Nabhan, G. P., & Trimble, S. (1994). *The geography of childhood*. Boston: Beacon.

Naherniak, C. (1995). Profound encounters: Classroom animals–more than responsible pet care. *Clearing, 91*, 12–15. ERIC EJ540104.

Ogden, J., & Heimlich, J. (2009). Why focus on zoo and aquarium education? *Zoo Biology, 28*(5), 357–360.

Patrick, P. (2010, September). *What middle school students know about zoos*. Poster presented at the annual meeting of the Association of Zoos and Aquariums, Houston, TX.

Patrick, P., Matthews, C., Ayers, D., & Tunnicliffe, S. (2007). Conservation and education: Prominent themes in mission zoo mission statements? *Journal of Environmental Education, 38*(3), 53–60.

Patrick, P., Matthews, C., Tunnicliffe, S., & Ayers, D. (2007). Mission statements of AZA accredited zoos: Do they say what we think they say? *International Zoo News, 54*(2), 90–98.

Patrick, P., & Tunnicliffe, S. (2009, September). *Zoo acuity model*. Poster session presented at the annual meeting of the Association of Zoos and Aquariums, Portland, OR.

Patrick, P., & Tunnicliffe, S. (2011). What plants and animals do early childhood and primary students' name? Where do they see them? *Journal of Science Education and Technology, 20*(5), 630–642.

Price, E., Ashmore, L., & McGivern, A. (1994). Reactions of zoo visitors to free-ranging monkeys. *Zoo Biology, 13*, 355–373.

Resnick, L. B. (1991). Shared cognition: Thinking as social practice. In L. B. Resnick, J. M. Levine, & S. D. Teasley (Eds.), *Perspectives on socially shared cognition* (pp. 1–20). Washington, DC: American Psychological Association.

Ross, S., & Gillespie, K. (2009). Influences on visitor behavior at a modern immersive zoo exhibit. *Zoo Biology, 28*, 462–472.

Shepard, P. (1982). *Nature and madness*. San Francisco: Sierra Club Books.

Simmons, D. (1994). Urban children's preferences for nature: Lessons for environmental education. *Children's Environments Quarterly, 11*, 194–203.

Smith, E. (1991). A conceptual change model of learning science. In S. Glynn, R. Yeany, & B. Britton (Eds.), *The psychology of learning science* (pp. 43–63). Hillsdale, NJ: Erlbaum.

Smith, L., Curtis, J., & Van Dijk, P. (2010). What the zoo should ask: The visitor perspective on pro-wildlife behavior attributes. *Curator: The Museum Journal, 53*(3), 339–357.

Stevens, B., Sams, K., & Ogden, J. (2004). *Catalyzing conservation: The backyard approach.* Presented at the Proceedings of the Catalysts for Conservation: A Direction for Zoos in the 21st Century. London Zoo: Zoological Society of London.

Surinova, M. (1971). An analysis of popularity of animals. *International Zoo Yearbook, 11*(1), 165–167.

Swanagan, J. (2000). Factors influencing zoo visitors' conservation attitudes and behavior. *The Journal of Environmental Education, 31*(4), 26–31.

Thurston, B. (1995, January 1). Zoo-do economics. *Pittsburgh Business Times*, p. 5.

Tunncliffe, S., & Reiss, M. (1999). Building a model of the environment: How do children see animals? *Journal of Biological Education, 33*, 142–148.

Tunnicliffe, S. D. (1996a). Conversations within primary school parties visiting animal specimens in a museum and zoo. *Journal of Biological Education, 30*(2), 130–141.

Tunnicliffe, S. D. (1996b). A comparison of conversations of primary school groups at animated, preserved and live animal specimens. *Journal of Biological Education, 30*(3), 1–12.

Tunnicliffe, S. D. (1996c). The relationship between pupil's ages and the content of conversations generated at three types of animal exhibits. *Research in Science Education, 26*(4), 461–480.

Tunnicliffe, S. D. (1997). The effect of the presence of two adults – Chaperones or teachers – on the content of the conversations of primary school groups during school visits to a Natural History Museum. *Journal of Elementary Science Education, 9*(1), 49–64.

Tunnicliffe, S. (2000). What sense do children make of three-dimensional, life sized representations of animals? *School Science and Mathematics, 100*(3), 1–11.

Tunnicliffe, S., & Reiss, M. (2000). Building a model of the environment: How do children see plants? *Journal of Biological Education, 34*(4), 172–179.

Turley, S. K. (1999). Conservation and tourism in the traditional UK zoo. *The Journal of Tourism Studies, 10*(2), 2–13.

Turley, S. K. (2001). Children and demand for recreational experiences: The case of zoos. *Leisure Studies, 20*(1), 1–18.

Vosniadou, S. (1998). *Cognitive psychology*. Athens, Greece: Gutenberg Publications.

Walker, S. (1991). *Education and training in captive animal management*. Proceedings in Perspectives in Zoo Management, National Zoological Park, New Delhi Zoo Ed Book, Zoo Outreach Organization.

Ward, P. I., Mosberger, N., Kistler, C., & Fischer, O. (1998). The relationship between popularity and body size in zoo animals. *Conservation Biology, 12*, 1408–1411.

Westervelt, M., & Lewellyn, L. (1985). *The beliefs and behaviors of fifth and sixth grade students regarding non-domestic animals*. Fish and Wildlife Service, United States Department of the Interior.

Wilkinson, B. (1997). *Multimedia wildlife education and attitudes*. Unpublished master's thesis, California State University, Northridge, CA.

Wineman, J., Piper, C., & Maple, T. (1996). Zoos in transition: Enriching conservation education for a new generation. *Curator: The Museum Journal, 39*(2), 94–107.

Young, I. (2008). *Mental models: Aligning design strategy with human behavior*. Brooklyn, NY: Rosenfeld Media.

Chapter 5
Exhibit Design

A zoo for the twenty-first century will contain elements of the conservation park, biopark, and theme park. The public hunger for amusement, not education, is the engine that will power the education and conservation programs. The visitor-friendly, unsophisticated theme attractions will be used to give scientific relevance to a fun experience, without being stuffy, preachy, or guilt-motivated. The emphasis will be upon solutions and personal involvement. Advanced technology not only will allow visitors to carry electronic pocket nature guides but also will provide invisible barriers between people and living animals. The animals, for their part, will move around through a realistic equivalent of home ranges, rotating through other species (Coe, 1996, p.115).

Exhibits

The first step in engaging a visitor is attracting them to an exhibit. The term exhibit is taken from the museum world where it is used to specify a stand-alone object and the display of an object within a setting. "Shettel likes to think of exhibits as 'idea generators', a term that captures the finest possibilities inherent in exhibits (O'Brien & Wetzelk, 1992, p. 5)." An exhibit is displayed by one party for the purpose of being viewed by another (Falk, Balling, Dierking, & Dreblow, 1985; Hensel, 1987). However, Miles, Alt, Gosling, Lewis, and Tout (1988) point out that

> ... objects by themselves, can communicate little beyond their own existence. The lesson of the exhibit designer must be that unless he wishes to restrict himself to an elite audience of scholars, who already know the background information, he must present his objects in a coherent and informative context. (p. 9)

In the case of zoos, the enclosure is the "setting" for the animals which are "actors," and the actors and setting are essential to tell the institution's story (Andersen, 1987; Coe, 1994). There are many types of exhibits in the museum world, but in zoos, exhibit refers to a group of animals displayed together with a linking theme. Hierarchically, the term *animal specimen* is the fundamental unit of an exhibit. Preserved *animal specimens*, such as skeletons and preserved organisms,

P.G. Patrick and S.D. Tunnicliffe, *Zoo Talk*, DOI 10.1007/978-94-007-4863-7_5,
© Springer Science+Business Media Dordrecht 2013

Fig. 5.1 Hierarchy of exhibit
terms

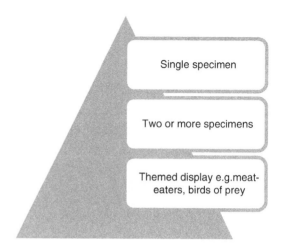

make up static exhibits. Live *animal specimens*, like those seen at a zoo, are part of dynamic exhibits. Exhibits may also incorporate two or more specimens within an exhibit. Several live *animal specimens* may be associated with other displays to form a theme (Fig. 5.1). The *animal specimens* and the exhibits are the key to the Zoo Voice and reflect the zoo's message. In fact, the

> Exhibits should fulfill three goals: to attract visitors, to hold their attention, and to communicate. Exhibits can communicate two types of messages: cognitive messages, having to do with concepts, knowledge, and information, and affective messages, relating to attitudes, feelings, and interests. (O'Brien & Wetzelk, 1992, p. 5)

Therefore, the zoo must consider how the exhibit expresses the messages they wish to convey. Exhibit design encourages visitors to look at the objects being displayed and to note important features through the animal being exhibited, the nature of the enclosure (i.e., immersion), the interpretation provided, and the emotive or hidden messages.

In the twenty-first century, zoo exhibits

> ... include a focus on multiple audiences, cultural interpretation, the use of storytelling as a technique, and having exhibits staffed with interpreters. Aquarium and zoo exhibits are becoming much more interactive, using hands-on experiences and animal encounters. We are shifting to a more issues-based approach to content and are discussing conservation at the ecosystem level. Aquariums and zoos are interpreting conservation success stories and promoting specific conservation actions, recognizing that audiences look to us for this leadership. The use of technology is evolving to include touch screens, cell phone tours and apps. (O'Connor, 2010, p. 4)

The second step in engaging visitors is to initiate visitor–exhibit interactions (Fig. 5.2). Visitor–exhibit interactions come in four forms: (1) walk past, minimum interaction or ignore exhibit; (2) passing comment, visitor makes a comment on some feature with a slight interaction; (3) explore, visitor stops, interprets the features, and has direct interaction with the exhibit by listening, talking, and/or

Category of behavior	Action	Interaction with exhibit
Walk past	Minimum interaction or ignore	No interaction
Passing comment	Walk by but remark on some feature	Slight interaction, some use of message displayed
Explore	Stop and interpret features using own experience and that of experts	Direct interaction by talking, touching, and seeking a response; aware of noise and smell of exhibit
Study	Stop and interpret using the message of the exhibit, everyday experience, and expert knowledge	Mental interaction— show and tell, re enact or teachback

Fig. 5.2 The interactions that take place within exhibits (McManus, 1987)

touching; and (4) visitor stops, interprets the exhibit using labels or uses their own expert/everyday knowledge, which leads to mental interactions such as show and tell, reenacting, or teaching the group (McManus, 1987). For zoos to become progressive in their attempt to bring about visitor–exhibit interactions, they may no longer view the zoo exhibit as just an animal in an enclosure. However, visitors do not always recognize the exhibit's message. Family interactions within exhibits indicate that exhibits have multiple *entry points* or multiple ways to understand what is essentially the same concept. One way to accomplish a multiple entry exhibit is to create thematic or cluster exhibits that concentrate on the big picture of science. Zoos could design their interpretive material and educational information based on the visitors' knowledge level and aid the visitor in constructing zoological meaning. Children, especially, are not spontaneously interested in the organisms' diet or its geographical origin. Visitors' cognitive ability, perceptual processes, and prior knowledge act as filters that may reduce the accuracy of the exhibit message, prevent the message from getting through, and distort the meaning of the exhibit (McManus, 1987). Visitors tend to look at, examine, and notice the characteristics with which they are familiar and do not cognitively challenge them.

Zoos are addressing differences in cognition and learning styles by offering visitors auditory, tactile, and visual opportunities to interact within an exhibit. Zoos provide opportunities for visitors to listen to oral presentations by offering key books where visitors insert a key into a machine to hear about the animal, handheld audio guides, and increasingly the use of visitor's mobile phones. Mobile phones may be used to listen to commentaries, navigate the zoo, and download maps and information (i.e., Houston Zoo). Moreover, zoos are recognizing that learning differs for each age group. An exhibit should appeal to the visitor's emotions by skillfully narrating a conservation story and providing information about particular animals and conservation research. For example, zoos are featuring specific exhibits such as the gorilla exhibit at the London Zoo. Zoo management recognizes that the message of the exhibit must intertwine the animal, the surroundings, the encounter, and a personal message from the zookeepers or docent volunteers. The interactions may be planned by the zoo staff but often appear casual and unplanned. These interpretive interactions with staff give visitors a feeling of being involved in the life and work of the zoo. Interpretation by curators and keepers is increasingly provided in zoos so that the visitor feels involved, which enhances their personal interpretation of the visit. However, some keeper interpretation does appear planned and provides specific information. For example, "Animals in Action" at the London Zoo is given by keepers and varies between fact giving and question answering with different outcomes for each visitor session (Visscher, Snider, & Stoep, 2009). In addition to the interactions that take place at zoos, many zoos have social media sites such as blogs, Facebook, and Twitter and post regular e-newsletters that extend the experience of the visit.

Visitors are more likely to comment about exhibits that are memorable, because they are unusual in design, elicit the senses, or are personally relevant. Verbal interactions may be influenced by a sense or a combination of the senses, because they elicit visitors' responses (Dale, 1954). The senses that are available may be real or simulated. Real sensory opportunities include touching real animal skins, hearing animal noises, seeing the animal, and smelling the animal. Noisy animals trigger visitor interest and hold the visitor within the exhibit longer (Patterson & Bitgood, 1987). However, visitors will comment on smells when they do not make other comments. For example, an 8-year-old child walking through the rhinoceros exhibit at the London Zoo commented on the smell saying, "Ah! What a whiff!" Simulated sensory opportunities occur when the visitor's safety is a concern or when the sensory opportunities would occur on rare occasions. For example, an animal's sounds may be broadcast throughout the exhibit or a QR code may be scanned to see a video of the animal. Of course, directly touching an organism is more likely to elicit a response than any of the other senses and seeing an animal educes the least amount of responses (Dale, 1954). Exhibits that incorporate multiple sensory stimulations, such as the hornbill exhibit at the Dallas Zoo, hold visitors attention longer than other exhibits (Koran, Koran, & Longino, 1986; Peart, 1984; Peart & Kool, 1988). For example, during a visit to the London Zoo, a grandmother used the senses of smell and sight to tell her 8-year-old grandson: "Here's the penguins! Feeding time is 2:30. Don't they smell?" The opportunity to experience an animal with more than one sense encourages visitors to spend more time in an exhibit observing and talking about the organism(s) being displayed. However, when the visitors run out of sensory interactions and conversational topics at an exhibit, they move on to another exhibit (Bitgood, Benefield, Patterson, & Nabors, 1986). The longer a visitor spends in an exhibit, the more likely they are to hear the Zoo Voice and leave with a message.

Exhibitry is progressing into a space that includes aspects of modern technology. New exhibit forms are designed around a conventional view and provide a message in a wider context than traditional exhibitry, which offered the organism's identity, diet, and geographical origin. Today's exhibits provide the visitor with real-time data and invite visitors to make observations, add to data collection, present evidence, and design informative posters. For example, when the Cat House at the Cincinnati Zoo and Botanic Gardens was upgraded, the new Night Hunters consists of a multisensory journey that embraced sight, sound, smell, and technology. Night Hunters exploits the senses to impose an emotive response as visitors traverse seemingly through the night. In addition to the clouded leopard, Pallas' cats, sand cat, fishing cat, caracal, and the black-footed cat, the exhibit features other nocturnal animals such as the Eurasian eagle-owl, potto, vampire bats, fennec fox, aardvark, and Burmese python. In Night Hunters, feeding tubes are located within the walls of the exhibit. These feeding tubes permit zookeepers to send live treats to the animals during the day, which allows visitors to observe the animal's feeding behavior. During a visit to Night Hunters, visitors can do Wild Research and create a poster that describes how they can help to save wildlife. When the visitor finishes the poster, they can email the poster to themselves or a friend. The signage within the exhibit is also evolving into the forward realm of technology as zoos like San Diego Zoo, Pittsburgh Zoo, and the Santa Barbara Zoo embed QR codes within the exhibit. The QR codes are being utilized by visitors to download information about animals. The San Diego Zoo provides a scan for Stickybits, which is a free phone application that allows visitors to share their pictures, videos, audio, and ideas with other animal enthusiasts.

In opposition to the new age of technology, the Condor Room at the Los Angeles Zoo contains no signage or guided activities but tells a unique story. The concern for developers of this untraditional exhibit is that people will not understand the function and the message. During exhibit design, developers should keep in mind that the message being transmitted is from the zoo's understanding of the concepts not from the visitor's perspective. However, the children, who visit the Condor Room, spend hours interacting with the exhibit. The Condor Room is an excellent example of a truthful, authentic exhibit that tells a unique story and translates the story for the visitor. Moreover, the emotive aspects of the Condor Room target the hearts and minds of the visitor by developing a conservation message that promotes action and education. In addition to exhibits with no signage, zoos are becoming increasingly aware of the importance of natural play areas (see also Chap. 2). Examples include the following: (1) The North Carolina Zoo Garden Friends Playground and the Henry Vilas Zoo Tree House and Adventure Play Area are traditional play areas with imaginative sculptures that provide children with unique places to climb, crawl, swing, and jump. (2) The North Carolina Zoo kidZone provides rope walks and outdoor activities for children. (3) The Calgary Zoo's Schoolyard Naturalization program assists schools in changing their school grounds to outside classrooms. (4) The San Antonio Zoo Kronkosky's Tiny Tot Nature Spot where children might feed lettuce to a tortoise or dig up earthworms. In this child-centered exhibit, children touch animals, climb, dig, and get wet and dirty as they learn about the natural world.

The Zoo Voice shapes the awareness of visitors and impinges on the sociocultural knowledge of the visitor. As the Zoo Voice seeks to bridge the gap between visitors and their knowledge of the natural world (Louv, 2006) and the natural history of a species, the zoo incorporates elements within their exhibit designs that bring visitors closer to the animals and reflect sociocultural viewpoints. Bringing visitors closer to the animals is achieved by using low-level camera traps in the same way a field conservation biologist uses the camera traps to locate wild animals. Exhibits, in which glass separates the visitor from the animal, provide a close-up opportunity for engaging the hearts and minds of visitors. Heated rocks are used to entice animals to sit near the glass, where visitors can place their hand against the paw of a polar bear or lie alongside a leopard or lion. Using the sense of smell is increasingly being used to involve the visitor in the exhibit. Sniff holes, located in the polar bear exhibit at the Columbus Zoo, allow visitors an opportunity to smell a polar bear. Peep holes or underwater observation areas allow visitors to feel like they are viewing the secrets of the animals. The North Carolina Zoo Rocky Coast exhibit and the London Zoo Penguin exhibit are excellent examples of allowing visitors to see the unseen life of the organisms. These seemingly hand-to-paw interactions between visitors and the organisms were not as readily available in the past as they are today (Anderson, 2003). Moreover, the exhibit design enhances the zoo visit through its inclusion of sociocultural artifacts. People view, make sense of, and identify an organism within the context of an exhibit and give the organism meaning within their own cultural perspective (De Witt & Osborne, 2007). The Malaysian House at the Twycross Zoo and the Audubon Zoo Sri Lankan Asian Elephant Exhibit Orientation Station have authentic Malaysian and Sri Lankan artifacts. The Tibetan prayer flags and native plants in the leopard exhibit at the Central Park Zoo introduce a cultural aspect to the exhibit. However, the exhibits themselves are not the only sociocultural opportunities provided by zoos. The Woodland Park Zoo's most visited webpage is Zoo Tunes. Zoo Tunes is a summer concert series that is sponsored by the Woodland Park Zoo (Bennett, 2011). Sociocultural artifacts and opportunities are an integral part of learning about the human society that shares geographical space with the organism, but the inclusion of sociocultural artifacts may distract some visitors from learning about the organism.

Dynamic exhibits are participatory and some offer controlled animal feeding opportunities, which place the visitor within the physical reach of the organism. Dynamic exhibits will promote conservation biology by providing realistic experiences.

> Carnivores will even simulate stalking, hunting, killing, and devouring their prey. You'll get the same thrills as if your family witnessed these events while on a foot safari in Samburu National Park. Or you'll be able to sit quietly among a troop of gorillas as ecotourists have done in Rwanda. After this experience, conservation biology and ethology will really mean something! Visitors will be able to sleep in tented camps, watch through night lenses at rhino and leopard come to the water hole, and go on sunrise bird and game walks. You can push your raft across a river just as a herd of elephants lumbers up. Or you can get a close look into a tunnel of harvest ants, and through a link-up between virtual reality and real-time micro video, you can venture into their tunnels with them. Adventure and learning

will be linked, as they were in the beginning—when your ancestors and mine climbed out of their trees at sunrise and set off across that ancient savanna in search of food. It is a link we would do well to rediscover. (Coe, 1996, p. 115)

Dynamic-passive man-made exhibits within zoos are composed of animatronics, such as the dinosaur. Animatronics are not normally avowed as the bastions of hands-on interaction that are afforded to other dynamic exhibits. Increasingly, animatronic exhibits are being utilized to model the prehistoric relatives of zoo animals. However, if animatronics are not integrated into a narrative exhibit, they tend to be ignored by children except for a quick glance. While observing animatronics, many visitors notice the wiring, the creaking of the machinery, and the joints in the body coverings, but not the image of the animal represented (Tunnicliffe, 1999). The Houston Zoo has recently installed animatronic dinosaurs in a grove of trees, where they offer an interesting vista for visitor. One of the dinosaurs even ejects fluid from its mouth at intervals and therefore promotes some interaction with visitors, who were dodging the jet. The responses at the dinosaur exhibit were of amazement, and visitors named familiar species (Scheersoi & Tunnicliffe, 2010).

No matter which exhibit design the zoo uses as a voice for its story, exhibit designers strive to present the nature of science or science as a way of knowing by including multiple viewpoints, scientific expertise, personal values, and ethical perspectives (McCallie et al., 2009). The exhibit provides information concerning how humans have impinged on the fauna and flora of the world and defines some of the greatest issues facing organisms. For instance, the Monterey Bay Aquarium has installed an exhibit that explains the difference between bleached dead coral and active coral and the possible causes of coral death. In addition to their work within the aquarium, the Monterey Bay Aquarium has successfully pioneered seafood cards that alert the public to sustainable seafood and which they are encouraged to leave in restaurants.

Zoo visitors have the potential to interact with exhibits in a number of ways that allow them to become mentally and cognitively engaged. Using their personal context, prior knowledge, and the information provided by the institution, visitors explore an exhibit through their mental interactions. Visitors' mental interactions range from observation and discussion to constructing new meaning and acquiring new understandings. Visitors may relay information to their group and share prior knowledge or new information gleaned from the exhibit. Reading a label and sharing information passes on information to the group. A successful exhibit, from the institution's perspective, is one that entertains, educates, and transmits a message that is received and comprehended by the visitors. Institutions may consider themselves to be successful in conveying their conservation and biological conservation missions if they perceive visitors as having conversations within the exhibits (Screven, 1986). However, Hancocks (2001) states that a successful zoo visit may be judged by two criteria: (1) how the audience has perceived the experience of interacting with the exhibit and (2) the extent that the message of the institution has been received and understood.

Successful institutions do not only rely solely on conversations; successful zoos design exhibits and interpretive techniques that link what the visitor already knows and feels with new information. However, identifying the topics of conversations visitors share and contribute during a zoo visit, we may be able to determine the link between the Zoo Voice and the Visitor Voice. Only by identifying all the pertinent parts of a zoo visit will a meaningful "museum experience" (Falk & Dierking, 1992) be created for the visitor in terms of personal context, enjoyment, and the acquisition of novel information.

Labels

Labels are an integral part of exhibits and the Zoo Voice. The exhibit label is utilized more often than any other method to communicate with zoo visitors, and visitors rely on labels for pertinent information (Jacobi & Poli, 1995; McManus, 1987, 1989, 1990, 1991). During a zoo visit, labels are the main source of information to which visitors refer in the course of a conversation (Tunnicliffe, 1995). The exhibit labels provide visitors with information about the exhibits, facilities, and the zoo's conservation mission. Moreover, the information incorporated in a zoo exhibit tells the story of the object being exhibited (Greenglass, 1986; Pearce, 1992; Ravelli, 2006). The object's story can be interpreted at three levels: (1) everyday, (2) informed, or (3) expert.

Different types of exhibits elicit different patterns of label reading behaviors (Rosenfeld, 1980). Visitors are not required to read the labels that are provided at exhibits. If the label is difficult to read or does not match the visitors' expectations of pertinent information, the visitor will not use them (Screven, 1992; Thompson, 1990). When an object does not have a label, the object is considered "silent." In other words, the institution has nothing to say about the object being displayed. However, the unlabeled object may be interpreted by the visitor using their personal knowledge and understanding. Weiner (1963) believes that the information provided to the visitor is only as good as the information's ability to communicate. If the words in the exhibit are not easily decoded or understood by the visitor, then the visitor may as well be standing in an exhibit that does not include labels. The label, as a representative of the Zoo Voice, is a catalyst for zoo visitors' conversations (Visitor Voice) (McManus, 1987, 1989; Screven, 1992; Serrell, 1981). If visitors are not able to use the labels to identify objects, then the label has not upheld its critical function of information transmitter.

There are exhibits that do not provide labeling, such as the previously mentioned Condor Room at the Los Angeles Zoo and drive-through safari parks. The visitor is compelled to use their own knowledge and understanding of the object on display. Establishments which provide exhibits with no overt labels expect the visitors to use their own interpretation. The lack of factual information within an exhibit may focus the attention of the visitors on affective aspects of the animals. Conversely, labels in live collections may be an integral part of the exhibit but may be ignored by the visitors. The visitors may be attracted to the affective domain of the exhibit and the organisms being displayed instead of to the labels. The labels at exhibits, in which animals are moving, are not used in a similar manner to those at other types of exhibits. Visitors locate and observe the animal, rather than reading the label that is located within the exhibit. In this case, the label is read after the specimen has been viewed (Serrell, 1981).

Labels and their location affect the amount of time visitors spend reading labels (Diamond, 1986; Screven, 1992) and the conversations that occur at the exhibit. Moreover, the length and conceptual complexity of the label's script influence how long visitors spend within an exhibit (Riddle, 1980). For example, some labels take less time to read, while other labels may provide more information or pose questions (McIntosh, 1993). However, sometimes the labels may only contain support material and not information concerning the actual animals. Hensel (1987) has found that aquarium visitors point to the labels more than they point to the actual animals. This is because visitors appear to be less familiar with aquarium animals than zoo animals. Additionally, different groups of visitors use labels or signage in varying ways. Nonchildren groups spend more of their visit engaged with signage than groups with children. Nonchildren groups spend more of their social interaction time with each other and spend less time engaging with signage (Ross & Gillespie, 2009). Therefore, the number of labels, clarity of the labels, the information provided on the labels, and the group members may influence the visitors' learning experience and the conversations they have during and after the visit.

Listening to visitors conversations within zoo exhibits reveals that visitors use labels to (1) name the animal, (2) identify the animal's attributes as a species and as an individual (e.g., age, births, sex, weight, and name such as the Panda bear Yang Yang at the Atlanta Zoo), (3) correct a visitors misidentification of an animal (such as a child identifying a dog and the parent looking at the label to determine the animal is a gray fox), (4) find management information, (5) learn about conservation issues and initiatives, and (6) discover current scientific field research (such as the effects of climate change on the species) (Hensel, 1987; Rosenfeld, 1980; Taylor, 1986). During a fieldtrip, teachers or chaperones are likely to fill the role of label reader, but students may also be encouraged to read the labels for the group as part of the educational experience. During a family visit, a person takes on the role of label reader (Hensel, 1987; Taylor, 1986). Hensel has recorded the following conversation during a family's visit to an aquarium:

Child:	Fishes.
Child 2:	What's that?
Adult female:	Jennifer, did you look at the sea horse? Oh! Come here Rob.
Adult female:	A blue faced angel fish. [reading]
Child:	Blue faced angel fish. [points to exhibit, then adults walk away, children follow]

Animals as Exhibits and Topics of Conversation

The contents of the exhibit influence the attention that visitors pay to the exhibit and hence may induce personal learning experiences. As a zoo specimen, the animal becomes an exhibit, takes on the "mantle of history," and becomes a part of the Zoo Voice. The animal becomes the educational tool of the institution. Animals as exhibits fit into the overall pattern of exhibit theory as is shown in Fig. 5.3. Animals are part of dynamic exhibits, and visitors may interact with them passively by walking by or actively by trying to initiate contact with the animals. Animal contact may also be instigated by the staff, other people in the exhibit, or the interactive elements or activities provided within the exhibit. Living animals that are more animated elicit more diverse and long-lasting conversations Tunnicliffe (1996). Nonliving animals that move, such as animatronics, educe more conversations with deeper content when they are integrated into an exhibit that provides evolutionary context. Animatronics that are solely used as moving animal displays without a narrative do not engage the visitor (Tunnicliffe, 1999). The animals in the exhibit, living or preserved, contain their own messages that are inherent in their structure and attributes and are interpreted by the visitors whether they or labeled or not. The interactions visitors have while visiting an animal exhibit and the experience(s) they take away from the visit are shaped by (1) the visitors' preexisting attitudes toward animals, knowledge of animals, and experiences with animals; (2) the emotions the exhibit arouses in the visitor; (3) the senses the visitor uses while experiencing the exhibit; (4) the animals' visual impact; (5) the reactions of the other member of the visiting group; (6) whether the animals are living, nonliving, or preserved; (7) the number of specimens within an exhibit and the visitors ability to see them; (8) the psychological involvement required by the exhibit; and (9) the conversations that take within an exhibit.

When visitors are in an exhibit, they may have verbal interactions by commenting on the exhibit, reading the signage and teaching others about the organism(s) (McManus, 1987), or talking directly to the organism (Rosenfeld, 1980). Visitors do talk directly to animals (Hensel, 1987) as shown in this short dialog at a penguin exhibit that takes place between a 4-year-old and his grandmother:

Neil: Hello! Hello!
Grandma: Aren't they cute?!
Neil: Hello! Hello!

If the visitor perceives that the organism's behavior is altered or the organism produces a sound, the visitor's mental model will identify this moment as a memorable experience that becomes part of their conversations. In addition to animal comments, visitors are also concerned with bathrooms, food, gift shops, and the location of the next exhibit. These verbal interactions are management comments and may be directed at members of the group or at an individual. Looking at features of the exhibit accounts for over half of visitors' comments, but direct reference to labels is not a prominent part of the conversations.

static exhibits				dynamic exhibits							
Physically passive watch, listen, smell		Physically active touch, manipulate		Physically passive watch, listen, smell				Physically active touch, manipulate			
Models	Preserved	Models	Preserved	Live	Constructed			Live	Constructed		
A few specimens in a museum	Most specimens in a museum	Statues in zoos. Some may be part of exhibits as an accessory like footprint casts	Some specimens found in a discovery center	Most live specimens	Passive	Active		A few opportunities in children's zoos at animal encounters	Passive	Active	
					Automatic models such as animated dinosaurs	Predictable outcomes	Unpredictable outcomes		Opportunistic involvement	Predictable outcomes	Unpredictable outcomes
						Visitor can choose not to be actively involved	Watches others			Push buttons or turn handles	Visitor is involved and may use computer programs

Fig. 5.3 Visitor participation within exhibit types based on Miles and Tout (1992)

The affective aspects of the physical context of zoos, such as a strange location and the use of living or preserved animal specimens, may influence the comments made by visitors. Moreover, the animal's behavior and the artifacts (droppings, tracks, odors, sounds, food remnants) the animal creates are important aspects of visitor learning. The type of living (zoos) and preserved (museums) specimens displayed influence what children learn and their memories of a field trip. Children who visit zoos generate far more affective responses to the living animals than children who see preserved specimens in museums (Birney, 1988). Zoo visitors may comment on an organism's body parts, behavior, color, structure, etc. The issue is that we do not know enough about visitor's reactions to or expectations of animal structures and behavior. Uncovering visitors' ideas of animals, prior knowledge, degree of perception, responses to an exhibit, and conversations that take place during a zoo visit will challenge and guide exhibit designers. Providing visitors with exhibits that defy their common patterns of conversation and response while viewing animals will assist the visitor in learning more about the specimens presented. Recently, exhibition developers and evaluators have sought to foster conversation among visitors as part of the exhibition experience (McLean & Pollock, 2007). Exhibits may spark bidirectional conversation, but in some exhibits, the presenters (zookeepers or docents) may prevent dialog among the visitors during the in exhibit experience (McCallie et al., 2009).

Of course, as previously mentioned the interactions and conversations that occur within a group are different based on the group's composition. Spontaneous conversations at animal exhibits among school groups and family groups reveal similarities in comments that name the animal and identify body parts and behaviors. However, families generate fewer "other exhibit" comments, such as smell and touch, than school groups. As previously mentioned, the type of animal exhibit (living, nonliving, and preserved) influences the structure of the visitors' conversations. At moving exhibits, such as animatronics, affective comments are generated by school groups, and behavioral comments are generated by both family groups and school groups. School groups' conversations at animatronics include more comments about behavior than their conversations at zoo and museum animals, but school groups' conversations at zoos and museums include behavioral comments at a similar rate. Families' conversations at zoos and animatronics include similar behavioral comments, but families' conversations at museums include the least amount of behavioral comments. The advantage of animatronics is that they move in a planned sequence repeating their narrative and children are able to predict the actions and behavior and comment on the moving anatomy.

Children in family groups not only comment about behavior and their observations of the animal, they interpret what they see by asking questions, making statements, and explaining their observations in terms of previous experiences. Children's comments are largely anthropomorphic in nature. Visitors express a range of affective comments often interpreting what they see in anthropomorphic terms. The following conversation between a mom and a 7-year-old at a toucan exhibit expresses the child's concern that toucan is alone (anthropomorphism):

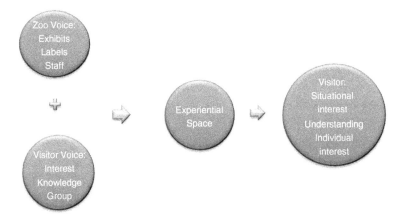

Fig. 5.4 The experiential space for learning that develops within a zoo exhibit

Michael: It's sad!
Mum: Why is it sad?
Michael: He's got no other birds.

Experiential Space in Exhibits

The Zoo Voice tells the story of the zoo, but the Visitor Voice may override or ignore the zoo's story (Fig. 5.4). The Zoo Voice must be compelling and become the hook that captures the visitors' interest. The space within the exhibit where the Zoo Voice captures the visitors' interest and the Visitor Voice reflects the Zoo Voice is an Experiential Space (Scheersoi & Tunnicliffe, 2010). During a zoo visit, the Zoo Voice provides the exhibits, the labeling, and the staff interactions. The Visitor Voice brings prior knowledge, interest in the exhibit, and the group with which the visitor is included. All of the interactions within and between the Visitor Voice and Zoo Voice have the potential to influence a visitor's learning. The zoo exhibit, labeling, and staff; a group member; interest in an exhibit or animal; and prior knowledge may shape the scaffolding outcome of a visitor's knowledge. Within the Experiential Space, visitors may develop situational interest, individual interest, (Scheersoi & Tunnicliffe), and personal understanding.

Situational interest refers to the condition and motivation that is related to a specific situation (Krapp, 1992, 1993; Mitchell, 1993; Renninger, Hoffmann, & Krapp, 1998; Todt & Schreiber, 1996). This book identifies zoos as the place where situational interest takes place. The synthesis of Bulunuz and Jarrett's (2010) theoretical framework may be used to explain the relationship between students, situational interest, and individual interest found within the experiential space within a zoo exhibit.

Working within the conceptual framework of Dewey (1913,1979), Krapp (2004) espoused creation of learning environments in which students actively interact with materials to reach an actualized state called *situational interest*, eventually developing into an enduring and more diffuse state, called *individual interest*. Krapp (2004) hypothesizes that transition from situational to individual interest can occur only if both feeling-related experiences and cognitively represented factors are experienced together and positively. Glasser (1998) describes fun and freedom to choose as basic human needs, suggesting that students in classrooms where science is fun and where there is student input might develop more interest in science. (Bulunuz & Jarrett, 2010, p. 66)

Situational and individual interest should be a concern of zoos, because they are believed to develop enduring value beliefs. Situational interest occurs, while the visitor is aroused or excited within the experiential space and may "facilitate specific motivations to act (Schiefele, 2009, 198)." Within the experiential space at the Durrell Wildlife Conservation Trust, visitors may come in contact with the Jersey Agile Frog and become concerned about its potentially terminal decline. Individual interest occurs when people have "a high level of interest in a particular subject area" that requires "close associations between that subject area and positive feeling- and value-related attributes (e.g., excitement)" (Schiefele, pp. 197–198). For example, when visitors leave the Durrell Wildlife Conservation Trust, they may ask the Agile Frog group to visit their home and help them change their lifestyle to better help the Jersey Agile Frog.

Even when the exhibit is fun and does work to engage the audience, situational interest may not occur. The interaction that occurs within a zoo exhibit, between the person and the object being viewed, may not elicit situational interest if an individual's interpretation does not "make sense" and they are not able to link the new information with an existing idea or mental model (Gilbert & Boulter, 2000). The realization of situational interest requires a situation-specific interaction between the person and the object. However, some visitors may not interact with another person but may only stand and stare or observe the exhibit. The result can be that the person has no interest or that interest emerges in response to situational cues (Krapp, 1999; Schiefele & Krapp, 1991). If situational interest does occur within the experiential space, it can deepen the zoo experience for the individual and cause a lasting individual interest in the experience. Because the experiential space is different in every exhibit, visitors' situational interest will vary from exhibit to exhibit.

Gaining the interest of visitors is a crucial component of the zoo's ability to impart information concerning conservation biology (Zoo Voice). Zoos must interest visitors not only in organisms but also in organism survival. To accomplish this task, zoos must overcome the physical barrier of the exhibit and understand visitors' learning to develop the experiential space between the visitors and the exhibit. However, families' conversations within an exhibit do not link or apply ideas from a previously visited exhibit (Parker, 2007). Therefore, each exhibit is a separate learning event and forms a distinct experiential space. The zoo's ability to transmit its story in its exhibit interpretation and develop the experiential space is not easy endeavor, because each visitor experience is different. During the zoo visit, the visitors' prior knowledge and interest in the subject matter and people within the visitors' group and within the exhibit may scaffold learning and assist visitors in constructing knowledge. Visitors will automatically label the specimen or activity (Bruner, Goodrow, & Austin, 1956), comment on and interpret behavior (e.g., Tunnicliffe, 1996), and identify specific features of the environment (Krapp, 1999; Schiefele & Krapp, 1991). The zoos' contribution to the experiential space is its specimens, habitats, staff, and cultural practice. The experiential space at an exhibit is the place where the zoos' contributions and the visitors' involvement find a common ground. The Zoo Voice and the Visitor Voice become the messenger and the receiver and promote the beginning of a multigenerational, bidirectional dialog. Formal and informal educators can have an enormous impact on the development of situational and individual interest by influencing the quality of interaction between the person and the object. However, this educational task is complex and challenging. Most visitors possess a limited amount of knowledge about conservation biology, which causes an understanding deficit during a zoo visit. Situational interest is easier to cultivate if the visitors are engaged, excited, and having fun. Excitement and interest are crucial in holding the visitors' attention and encouraging learning, because zoo visitors rarely visit with learning as their main objective.

The Zoo Voice inherent in the exhibit and enhanced and extended by interpretation such as labels, video clips, keeper talks, and interactive opportunities is important to the zoo. However, the Zoo Voice is but one side of the equation. The knowledge the visitor brings to the zoo is also a critical element but frequently seems misunderstood or under utilized by zoo staff, scientists, designers, and donors. The staff, scientist, designers, and donors each have a particular agenda when providing their input into the exhibit design. However, a clear, concise understanding of visitors' prior knowledge is essential, specifically their knowledge and understanding of organisms and the natural world.

References

Andersen, L. L. (1987). Right enclosure design- before stories can be told. In P. Haase (Ed.), *Education/interpretation – Trends for the future* (pp. 26–52). Copenhagen, Denmark: Copenhagen Zoo.

Anderson, D. (2003), Visitors' Long-term Memories of World Expositions. *Curator: The Museum Journal, 46*, 401–420.

Bennett, J. (2011, September 12–17). Summer concert series via the web site. *How to generate revenue through your web site*. Presentation at the Association of Zoos and Aquariums Conference, Atlanta, GA.

Birney, B. (1988). Brookfield zoo's flying walk exhibit: Formative evaluation aids in the development of an interactive exhibit in an informal learning setting. *Environment & Behavior, 20*(4), 416–434.

Bitgood, S., Benefield, A., Patterson, D., & Nabors, A. (1986). *Understanding your visitors: Ten factors that influence visitor behaviour*. Jacksonville, AL: Psychology Institute, Jacksonville State University.

Bruner, J. S., Goodnow, J. J., & Austin, G. A. (1956). *A Study of thinking*. New York: Wiley, Science Editions, Inc.

Bulunuz, M., & Jarrett, O. (2010). Developing an interest in science: Background experiences of preservice elementary teachers. *International Journal of Environmental & Science Education, 5*(1), 65–84.

Coe, J. C. (1994). Design and perception: Making the zoo experience real. *Zoo Biology, 4*(1), 197–208.

Coe, J. C. (1996). Future fusion: The twenty-first century zoo. In N. Richardson (Ed.), *Keepers of the kingdom: The new American zoo*. Charlottesville, VA: Thomasson-Grant & Locke.

Dale, E. (1954). *Audio-visual methods in teaching*. New York: The Dryden Press.

Dewey, J. (1913/1979). Interest and Effort in Education. In J. A. Boydston (Ed.), *The middle works, 1899–1924* (Vol. 7 1912–1914, pp. 153–197). Carbondale, IL: Southern Illinois University Press (Original work published in 1913).

DeWitt, J., & Osborne, J. F. (2007). Supporting teachers on science-focused school trips: Towards an integrated framework of theory and practice. *International Journal of Science Education, 29*(10), 685–710.

Diamond, J. (1986). The behavior of family groups in science museums. *Curator, 29*(2), 139–264.

Falk, J. H., Balling, J. D., Dierking, L., & Dreblow, L. (1985). Predicting visitor behaviour. *Curator, 28*, 249–257.

Falk, J. H., & Dierking, L. (1992). *The museum experience*. Washington, DC: Whalesback Books.

Gilbert, J. K., & Boulter, C. (2000). *Developing models in science education*. Dordrecht, The Netherlands: Kluwer.

Glasser, W. (1998). *Choice theory: A new psychology of personal freedom*. New York: Harper Perennial.

Greenglass, D. I. (1986). Learning from objects in a museum. *Curator, 29*(1), 53–66.

Hancocks, D. (2001). Is there a place in the world for zoos? In D. J. Salem & A. N. Rowan (Eds.), *The state of the animals 2001*. Washington, DC: Humane Society Press.

Hensel, K. (1987). *Families in Museums: Interactions and conversations at displays*. Unpublished Ph.D. thesis. Columbia University Teachers College, New York.

Jacobi, D., & Poli, M.S. (1995). Scriptovisual documents in exhibitions: Some theoretical guidelines. In A. Blais (Eds.), Text in the exhibition medium (pp. 33–46). Montreal Canada: Societe des Musees Quebecoi & Musee de la Civilisation.

Koran, J. J., Jr., Koran, M. L., & Longino, S. (1986). The relationship of age, sex, attention, and holding power with two types of science exhibits. *Curator, 29*(3), 227–235.

Krapp, A. (1992). Konzepte und forschungsansätze zur Analyse des zusammenhangs von interesse, lernen und leistung. In A. Krapp & M. Prenzel (Eds.), *Interesse, lernen, leistung* (pp. 9–52). Münster, Germany: Aschendorff Verlag.

Krapp, A. (1993). *The construct of interest-characteristics of individual interests and interest-related actions from the perspective of a person-object-theory*. München, Germany: Universität der Bundeswehr München.

Krapp, A. (1999). Interest, motivation and learning: An educational-psychological perspective. *European Journal of Psychology of Education, 14*(1), 23–40.

Krapp, A. (2004). Interest and human development: An educational-psychological perspective. *British Journal of Educational Psychology, Monograph Series II, Part 2* (Development and motivation), 57–84.

Louv, R. (2006). *Last child in the woods. Saving our children from nature-deficit disorder*. Chapel Hill, NC: Algonquin Books.

McCallie, E., Bell, L., Lohwater, T., Falk, J., Lehr, J. Lewenstein, B., Needham, C., & Wiehe, B. (2009). *Many experts, many audiences: Public engagement with science and informal science education*. Washington, DC: Center for Advancement of Informal Science Education (CAISE).

McIntosh, L. (1993). As it happens! Connecting visitors to the dynamics of the aquarium and the wild. *Journal of the International Association of Zoo Educators, 29*, 32–38.

McLean, K., & Pollock, W. (2007). *Visitor voices in museum exhibitions*. Washington, DC: Association of Science-Technology Centers Inc.

McManus, P. M. (1987). *Communications with and between visitors to a science museum.* Unpublished Ph.D. thesis. Chelsea College, University of London, London.

McManus, P. M. (1989). What people say and how they think in a science museum. *Heritage Interpretation.* London: Belhaven Press.

McManus, P. M. (1990). Watch your language! People do read labels. *ILVS Review: A Journal of Visitor Behaviour, 1*(2), 125–127.

McManus, P. M. (1991). *Making sense of exhibits. Museum Language.* Leicester, UK: University Press.

Miles, R. S., Alt, M. B., Gosling, D. C., Lewis, B. N., & Tout, A. F. (1988). *The Design of Educational Exhibits.* London, UK: Unwin Hyman.

Mitchell, M. (1993). Situational interest: Its multifaceted structure in the secondary school mathematics classroom. *Journal of Educational Psychology, 85*(3), 424–436.

O'Brien, M., & Wetzelk, J. (1992). What's visitor evaluation all about? Report on a workshop conducted at the Baltimore Aquarium & the Maryland Science Center. *Visitor Behavior, 7*(2), 5–10.

O'Connor, T. (2010). *Trends in zoo and aquarium exhibit interpretation.* Oregon Coast Aquarium. Available from http://www.izea.net/education/interpretation.htm

Parker, L. C. (2007). *The use of zoo exhibits by family groups to learn science.* Unpublished Thesis/Dissertation. Purdue University, West Lafayette, IN.

Patterson, D., & Bitgood, S. (1987). *Exhibit design with the visitor in mind.* Jacksonville, AL: Center for Social Design.

Pearce, S. M. (1992). *Museums objects and collections.* Leicester, UK: University Press.

Peart, B. (1984). Impact of exhibit type on knowledge gain, attitudes and behaviour. *Curator, 27*(3), 220–237.

Peart, B., & Kool, R. (1988). Analysis of a natural history exhibit: Are dioramas the answer? *The International Journal of Museum Management and Curatorship, 7,* 117–128.

Ravelli, L. J. (2006). *Museum texts: Communication frameworks.* New York, NY: Routledge.

Renninger, K. A., Hoffmann, L., & Krapp, A. (1998). Interest and gender: Issues of development and learning. In L. Hoffmann, A. Krapp, R. A. Renninger, & J. Baumert (Eds.), *Interest and learning. Proceedings of the Seeon Conference on Interest and Gender* (pp. 9–21). Kiel, Germany: Institute for Science Education at the University of Kiel (IPN).

Riddle, W. (1980). *A study of adolescent visitors to the human biology exhibition.* Unpublished Master's Thesis. Chelsea College, London.

Rosenfeld, S. (1980). *Informal learning in zoos: Naturalistic studies on family groups.* Unpublished Ph.D. Thesis. University of California, Berkeley, CA.

Ross, S. R., & Gillespie, K. (2009). Influences on visitor behavior at a modern immersive zoo exhibit. *Zoo Biology, 28,* 462–472.

Scheersoi A., & Tunnicliffe, S. (2010, July). *Is it dangerous? Interest of zoo visitors as a key to biological education.* Paper presented at the European Researchers in Didactics of Biology Conference, Braga, Portugal.

Schiefele, U. (2009). Engagement and disaffection as organizational constructs in the dynamics of motivational development. In K. R. Wentzel & A. Wigfield (Eds.), *Handbook of motivation at school* (Educational psychology handbook series, pp. 197–222). New York: Routledge.

Schiefele, U., & Krapp, A. (1991). *The effects of topic interest and cognitive characteristics on different indicators of free recall of expository text.* Paper presented at the annual meeting of the American Educational Research Association, Chicago.

Screven, C. G. (1986). Exhibitions and information centres: Some principles and approaches. *Curator, 29*(2), 109–138.

Screven, C. (1992). Motivating visitors to read labels. *ILVS Review, 2*(2), 183–214.

Serrell, B. (1981). Zoo label study at Brookfield Zoo. *International Zoo Year Book, 21,* 54–61.

Taylor, S. M. (1986). *Understanding processes of informal education: A naturalistic study of visitors to a public aquarium.* Unpublished Ph.D. thesis. University of California, Berkeley, CA.

Thompson, D. (1990). The effect of length, type size and proximity of interpretive signs on reading in a zoo. *Visitor Behaviour, 5*(1), 4–7.

Todt, E., & Schreiber, S. (1996, June). *Development of interest.* Paper presented at Seeon Conference on Interest and Gender, Seeon, Germany.

Tunnicliffe, S. D. (1995). The content of conversations about the body parts and behaviours of animals during elementary school visits to a zoo and implications for teachers organising field trips. *Journal of Elementary Science Education, 7*(1), 29–46.

Tunnicliffe, S. D. (1996). Conversations within primary school parties visiting animal specimens in a museum and zoo. *Journal of Biological Education, 30*(2), 130–141.

Tunnicliffe, S. D. (1999). It's the way you tell it! What conversations of elementary school groups tell us about the effectiveness of animatronics animal exhibits. *Journal of Elementary Science Education, 11*(1), 23–36.

Visscher, N., Snider, R., & Stoep, V. (2009). Comparative analysis of knowledge gain between interpretive and fact-only presentations at an animal training session: An exploratory study. *Zoo Biology, 28*(5), 488–495.

Weiner, G. (1963). Why Johnny can't read labels. *Curator, 1*(2), 143–156.

Chapter 6
Talking About Animals

One link between the learning of science and the use of language is the development, of a specialized vocabulary for representing concepts and describing processes. In addition, spoken language provides a familiar medium through which a child can describe their conceptions of phenomena in order that teachers assess a level of understanding (Ollerenshaw & Ritchie 1998). Children cannot be expected to bring to a task a well-developed capacity for reasoned dialogue. This is especially true for the kinds of discursive skills which are important for learning and practicing science: describing observations clearly, reasoning about causes and effects, posing precise questions, formulating hypotheses, critically examining competing explanations, summarizing results, and so on (Mercer, Dawes, Wegerif, & Sams, 2004, pp. 263–265).

Adults and children come to the zoo with ideas and images of animals and information and do have some general prior knowledge of animals. Therefore, zoos should enhance learning in a way that addresses visitors' misconceptions about animals by identifying the misconceptions and challenging the perceptions that are not in accordance with science. This is a difficult process if the zoo is not aware of the visitors' prior knowledge, especially the visitors' knowledge of animals. Visitors' prior knowledge may be identified within six topics: (1) understanding of the term animal, (2) comprehension of layman's taxonomy, (3) perceptions of animal behavior and anatomy, (4) curiosity about individual animals, (5) emotional connection to animals, and (6) cultural understandings of animals. An essential part of exposing visitors' prior knowledge is clarifying the connections between the visitors' knowledge and their unprompted comments. Listening to visitors conversations during the zoo visit and identifying the detail of the conversational topics will provide zoo educators and educational researchers with insight into the prior knowledge of zoo visitors. Visitors' comments come from their existing mental models of animals, are accessed during their observations at the exhibit, and are expressed during their explanations to others in the visiting group.

P.G. Patrick and S.D. Tunnicliffe, *Zoo Talk*, DOI 10.1007/978-94-007-4863-7_6,
© Springer Science+Business Media Dordrecht 2013

Taxonomy and the Term Animal

In order for adults and children to maximize learning during a zoo visit, they must have the appropriate knowledge and vocabulary. When teachers prepare students for a zoo visit, they must take into account the age of the children, the rationale for the visit, and the students' familiarity with the term "animal." Children need to be familiar with the critical attributes and behaviors of animals and how these attributes and behaviors can be used to group animals. An important aspect of understanding how children identify animals is identifying how scientists, society, and parents use the term animal and how the term is absorbed by children.

Zoological nomenclature, a system established by Carl Linnaeus in the late eighteenth century, is an agreed-upon method used to scientifically name organisms. Scientists use Linnaean nomenclature as a shorthand way of providing a detailed summary of the structural relationships among organisms. Each species has one name composed of two parts. For example, the domestic cat's (*Felis domesticus*) specific name is *domesticus*, which cannot stand alone, and its general name is *Felis*. The species are grouped into categories of ascending order until the phylum and animal kingdom categories are established, but the species name is the one used most often by biologists. Therefore, the name *Felis domesticus* briefly conveys to scientist information about the animal's anatomy, morphology, physiology, structures, behavior, and genetic connections to other animals. The use of Linnaean taxonomy requires a command of the relevant taxonomic language and an understanding of the taxonomic hierarchies. This system is at odds with the names used by the majority of zoo visitors (Hage, 1993; Hensel, 1987; Rosenfeld, 1980; Taylor, 1986).

Most zoo visitors are not aware of the taxonomic characteristics of common organisms and are not able to place them taxonomically. However, understanding the elemental characteristics of animal groups is important. How can visitors appreciate the complexity and intricacies of the web of life if they are not able to understand the differences between a frog and a bird? By being able to identify the characteristics of organisms, visitors may better understand organisms' physiological needs and their importance ecologically. Zoos are beginning to pay attention to this complex matter of understanding organisms and their taxonomic characteristics. Students on field trips may have an opportunity to discuss the taxonomy represented at the zoo, but most visitors do not have these discussions and do not have an understanding of the taxonomic language.

When zoologists classify animals, they use critical attributes to identify animal specimens and place them in appropriate taxonomic categories, for example, the body coverings of mammals (hair), birds (feathers), and reptiles (scales). The ability to scientifically identify organisms has become more important because of the reduction in global biodiversity and the need to pinpoint the organisms under threat (Crisci, McInerney, & McWethy, 1994). Even though genetic techniques, such as DNA fingerprinting, have been developed to establish genetic relationships among organisms, being able to visually identify organisms taxonomically is still essential.

Children glean information about animals from animal images (stuffed animals, toys, children's books, television, Internet, classroom decorations) (Atran & Medin, 2008), the culture, community, family, and friends, and through personal encounters with animals (Tunnicliffe, Gatt, Agius, & Pizzuto, 2008). Real interactions with animals and experience in the wild, such as playing in nature, are paramount to children's understanding of animals. Children's out-of-door experiences will allow them to develop a theory about natural phenomena (Driver, Squires, Rushworth, & Wood-Robinson, 1994) and taxonomic nomenclature before they experience any formal teaching. Because of the personal experiences children have with animals prior to formal teaching, children do notice the physical attributes of animals and are able to sort the animals in the world around them (Bruner, 1983).

Naming an animal starts in the early years. A prerequisite for using the word *animal* and understanding its meaning is that the person doing the naming is aware of an appropriate word to use as a *name*. The person using the word *animal* must be able to recognize an animal and have the vocabulary necessary to communicate their thoughts. Children quickly acquire a varied vocabulary (Child, 1985). However, children may understand *context* first and then gradually acquire the *relevant words* (Donaldson, 1978). For instance, a young child learns about the structures and behaviors of the family's pet cat before (s)he learns the words for these attributes. Naming nouns are the first words children learn (Brown, 1958; Macnamara, 1982). The sequence of learning nouns first may be because mothers start the sequence (Ninio, 1980; Ninio & Bruner, 1978), including names for animals. Mothers point to an object and name the object by providing a word then the child is expected to relate the word to that object each time. Because mothers are the early teachers, they act as a filter selecting not only the objects but also the type of name or word to be taught (Anglin, 1977; Markman, 1989). Animals comprise a large part of the point and name approach of mothers because children have animal books and toys and pets, and there are animals in the local environment. Therefore, it is not a surprise that the names of animals comprise a large part of children's vocabulary. When Anglin (1977) reorganized the categories of words Rinsland (1946) identified in the first vocabulary of children, he found that 36 of the 275 words children used were types of animals.

The word *animal* is a concept. Concepts are identified by other concepts that are deemed to relate to the original concept. The related concepts are known as exemplars (Smith & Medin, 1981). For example, the exemplars for an animal might be color and shape, but the exemplars are different for each person and for each object they observe. A child may recognize a dog as an animal because it is a family pet but may not recognize a bird in their yard as an animal because it does not look like the family dog. The ability to identify and name an animal occurs in stages from recognizing that something is an animal until the child learns the defining animal features (exemplars) and is able to identify species. The ability to name, identify, and separate animals based on characteristics is categorization. "The learning and utilization of categories represents one of the most elementary and general forms of cognition by which man adjusts to his environment" (Bruner, Goodnow, & Austin, 1956, p. 20). Moreover, the categories man uses to identify animals and their correlation may be invented or discovered (Bruner et al., 1956). Therefore, the preoccupation or instinctive response visitors have with finding or allocating a name or label to an animal is not surprising.

Because a theoretical framework for how people categorize animals and assign them names is not currently available, we base this book on the existing theoretical framework of concept attainment. Markman (1989) states that for a child to acquire a concept, the child needs (1) an analytic ability to separate the holistic view of an object into its component properties; (2) a defined hypothesis-testing system that when faced with new exemplars generates possible properties, evaluates the properties against new exemplars, and revises, rejects, or maintains a concept; and (3) an ability to use criteria to evaluate novel objects to determine whether they are members of a category.

Categories in general have *best examples*. *Best examples* of animals are usually seen in classrooms when teachers define animals by introducing caterpillars and butterflies, frogs, and megafauna such as lions, elephants, and polar bears. However, the *best examples* that represent a category are based on characteristics or exemplars that are "essentially a matter of both human experience and imagination" (Lakoff, 1987, p. 125). Therefore, if the characteristic or exemplar of a group member is nonnatural (e.g., beautiful), then these exemplars have to be enumerated for category membership and identified and grouped by humans. Humans assimilate these nonnatural exemplars and create correlations and impose the exemplars on the group. Even though these nonnatural categories exist and may be perceived by the layperson, the nonnatural categories are not identifiable and usable in everyday discussions. Therefore, unless the accepted scientific definition of an animal (which have been created by humans) is provided during a discussion, confusion can occur. If learners are to identify specific attributes or exemplars of the categories into which animals may be allocated, the learners must possess the appropriate vocabulary. Learners must be able to recognize the constituent parts required for membership in a category (Rosch & Mervis, 1975).

Humans categorize the world around them by linking and grouping discriminatorily different inanimate and animate objects. However, there is a difference in humans' ability to discriminate like groups. Laypeople are only able to discriminate between characteristics and groups at a superficial level, while specialists have a more profound understanding of the categories. Because of their multilayered understanding, specialists are able to discriminate further between the members of a category and are equipped to develop hierarchical taxonomic systems. For example, a skilled herpetologist can identify a snake as belonging to the infraorder Alethinophidia or Scolecophidia and the family and genus to which the specimen belongs. Conversely, the lay visitor is likely to recognize an organism as a snake, identify large specimens as pythons, name a rattlesnake and a cobra, and establish a snake hierarchy. In the following transcripts of a family's visit to a zoo, the children want to see a snake and are able to identify the snake based on their knowledge of the exemplars that identify a snake. However, the group does not possess the knowledge of the discriminatory characteristics needed to name the specific snake:

Neil: Have they got some snakes?
Michael: It's on that ledge.
Aunt: What is it?
Michael: It's a snake of some sort . . .

However, at the next two exhibits, the children do have knowledge of the discriminatory characteristics needed to provide the common names of the snakes. The children identify the organism as a snake (category) and recognize the subordinate category to which it belongs. At the viper exhibit, Michael immediately names the viper. At the cobra exhibit, Michael identifies the cobra but challenges his own identification because he cannot see the cobra's hood. The labeling within the exhibit illustrates a cobra with its hood extended, which occurs only if the cobra feels threatened:

Viper
Michael: Oh look! Yes! That's a viper, they live in pots . . .

Cobra
Michael: That's a cobra.
Neil: Yes, it's a cobra.
Michael: No, it's not a cobra.
Aunt: Why not?
Michael: Because it hasn't got things on it (pointing to the illustration of the cobra's hood).
Aunt: The things at the side of the neck.
Neil: You cannot see it all the time.

In the viper and cobra exhibit, Michael applies his knowledge of the exemplars of *snake* and concludes the organisms are snakes. Moreover, Michael and Neil extend their exemplars and categorize the cobra as a snake, even though the illustration does not match their knowledge. The *extension* of a concept is the ability to identify objects that are instances of that concept. For example, *cat, dog, elephant,* and *aardvark* are instances of the concept "animal." Michael and Neil extend their knowledge of *snake* and are able to accept that the viper and cobra are objects that represent examples of *snake*.

Understanding exemplars are an important part of identifying animals and related species. However, physical exemplars can lead to misunderstandings when they are not fully understood. In the following conversation, a child believes a rock hyrax is a rat, but the mother corrects the name and provides the correct name:

Rock Hyrax
Child: Look! It's a rat.
Mother: No. It's not a rat. The sign says it is a Rock Hyrax.
Child: What is that?
Mother: I do not know. [She reads more of the sign.] It says [sign]: "Can you believe elephants, not guinea pigs, are related to these small, furry animals?"

Child: They look like guinea pigs not elephants.
Mother: [As they are walking away.] Humph! They really got that wrong.

Because the mother could not reconcile the difference in the physical exemplars between the elephant and the rock hyrax, she assumed the zoo must be wrong. The sign does continue to explain the relationship between the rock hyrax and the elephant, but because the rock hyrax and the elephant are not visibly similar, the mother told her children the zoo was incorrect. This should have been a learning moment for the child and the mother but instead became an accusation against the zoo's integrity.

Similarly, a 4-year-old boy in a baboon exhibit identifies the baboon's reddish-gray fur as that of a squirrel. He is familiar with the squirrel because he sees squirrels in his everyday life:

Baboon
Boy: Look squirrel!
Mother: I am not so sure. [Looks at the label]. It's not a squirrel but a baboon.
Boy: It's a squirrel.

The problem occurs when the child is not able to adjust prior knowledge and still believes the baboon is a squirrel. The boy insists that the animal is a squirrel because the salient features he observes match his memory of a squirrel's fur. He had to categorize the new animal based on prior knowledge and to meet his satisfaction.

Children begin classifying objects by arranging them according to their increasing size (Piaget & Inhelder (1969). The tendency of young children to group objects by size is characteristic of the earliest stages in classification (Inhelder & Piaget, 1964). Moreover, children group animals according to themes like *red* or *big* and do not employ a taxonomic hierarchy (Markman & Seibert, 1976; Smiley & Brown, 1979). For example, during a zoo visit, a toddler remarked, "*Big* worm! A *big* worm, there!" An 8-year-old boy at the shark exhibit in the London Zoo Aquarium remarked, "They are sharks, because they are *long*...." Children may be attempting to classify organisms based on spatial recognition (exemplar). The appearance of the organism (exemplar) is also an important feature children use to name organisms. Younger primary children use exemplars to identify if an animal is an animal because they are not able to extract many features from the animal's form. Five-year-olds use appearance to place objects into categories, but by the age of nine, they understand that a name represents an object and that the object's characteristics could be inferred from the name (Carey, 1985; Keil, 1979; Markman, 1989). Additionally, 5-year-olds have the potential to organize categories hierarchically but with a reduced number of categories (Keil, 1979). Eleven-year-olds use exemplars and unique identifying characteristics when they group animals. Therefore, the greater the variety of animal forms that children see, the more they will practice grouping animals according to their different attributes. As children come in contact with more animals, children will learn basic animal names and begin to recognize the principal attributes (exemplars) of the categories (common or zoological) into which the individual organism belongs. Because young children are unable to form true abstract concepts until adolescence (Vygotsky, 1962), children should not be asked to examine and implement hierarchical taxonomies until they are 14 years old and may reason without concrete clues (Lowell, 1979).

Young children in the Piagetian concrete stage of learning need visual clues and are unlikely to be able to use a true taxonomic hierarchy. Even though young children are able to hierarchically divide objects, young children do not understand a true hierarchical taxonomy such as the labels used in traditional animal collections. Children could be taught to group animals within a taxonomic hierarchy by helping them identify the attributes within a taxonomic level. For example, children could be taught to name animals using their common name instead of a group name, for example, not "snake" but "python" and not "fish" but "shark." The role of classroom and zoo educators is to broaden the children's visual perceptions of animal diversity and provide examples of animal names and groups, with the goal of leading children to an understanding of zoological taxonomy. Children must learn the name of an animal, the category to which an individual animal belongs, taxonomic hierarchy of the category, and the animal's life stages. Children who are learning to use categorization encounter a number of problems. First, they have to learn whether the name being used by their companions refers to the whole object or some part of the object. For example, when mothers name objects for their children, they refer to the whole object not the objects' parts (Ninio, 1980). Second, animals have more than one name. A pet cat may have four names: a personal name (Sharma), an everyday name (cat), a common name (domestic cat), and a scientific name (*Felis domesticus*). Someone may say "There is a cat in the tree." and another person in the group may say "Sharma is in a tree." Third, an animal may be categorized as an individual animal using a basic name such as cat but also may be called a carnivore. A cat is a subordinate category of animal, but carnivore is a superordinate category that encompasses other organisms. Fourth, animals undergo metamorphosis which means they do not always have the same appearance during stages of their life cycle. Animals that undergo complete metamorphosis such as Lepidoptera and Amphibia represent difficulties for children. For example, to identify Lepidoptera, children must be familiar with the egg, larva, cocoon, and caterpillar. Sometimes, children view the tadpole and the frog as different representations of an animal. Eventually, young children understand the use of the superordinate category and the transitivity between categories and acquire an ability to view classes as individual inclusions within a hierarchy. Therefore, children will recognize that *if* an animal belongs to one subordinate category, it *must* also possess the defining features of the superordinate category that subsumes it (Markman, 1989). For example, children have difficulty distinguishing penguins as birds. Children believe that birds fly, but penguins swim, but once children understand that penguins have feathers (exemplar), then the children are able to transitively categorize penguins as birds.

Until children acquire the ability to understand the exemplars of a concept, class inclusion, and transitivity, children are likely to find the use of categorical names perplexing. Young children find focusing on more than one feature at a time difficult (Child, 1985). Inhelder and Piaget (1964) believe that young children develop hierarchical skills in three consecutive steps—Step 1: The spatial configurations of objects and their perceived similarities unite them into *graphic collections*. Several of the *graphic collections* are juxtaposed instead of being used as the basis of a hierarchical class structure. Step 2: *Nongraphic collections* are formed by children and show features of classification but no inclusion. Therefore, *all* members of a class are not recognized as belonging to their category (Piaget & Inhelder, 1969). Step 3: True hierarchical classification skills are acquired but are based on the child reaching the formal operational stage of cognitive development. Hierarchical classification requires abstract thought, uses logic, and involves embedded knowledge. However, hierarchical naming systems are not present in the ad hoc themes or collections of zoos (e.g., creepy crawlies) (Markman, 1989). The term *creepy crawlies* may be preferred by the children or the zoo, but the use of these simple terms does not assist children in developing an understanding of the zoological nomenclature and taxonomy of the organisms they see during a zoo visit.

Learning the categories and the appropriate terms for animals is an inductive task. Children must categorize the objects within the cultural context of their society, although theoretically, a number of different ways may be possible. Additionally, the child has to learn to what the label refers, identifying this from a myriad of possibilities. For example, mammals possess the salient characteristic of fur. Children intuitively want to know the names of things they see because the name is the code for a concept and enables the concept to be discussed (Markman, 1989). Understanding the ways in which people operate when confronted with unfamiliar species is important to both exhibit designers and zoo educators. Exhibit design and materials should help the visitors hierarchically categorize animals because the organism's environment consists of structures that provide visitors with connections to category construction (Gibson, 1979).

Ryman (1974) found that when 11-year-olds were asked to classify organisms based on characteristics, the children were not able to discern the organisms as a member or nonmember of a taxonomic group. These findings suggest that the children did not have the level of comprehension required to perform the task. Ryman pointed out that both inaccurate and imprecise concept formation and deficient language development contribute to children's difficulties in correctly classifying plants and animals accurately. These inaccuracies become a predicament for classroom teachers and zoo educators because when children identify animals, they apply their prior knowledge of animals' distinguishing features (Lucas, Linke, & Sedgwick, 1979; Mintzes, 1989; Sherwood, Rallis, & Stone, 1989; Tamir & Sever, 1980; Trowbridge & Mintzes, 1988). For example, older primary children list locomotory appendages (wings, legs, arms), body coverings, and attributes associated with the front end of the animals as distinguishing attributes (Bell, 1981; Braund, 1991; Mintzes, Trowbridge, Arnaudin, & Wandersee, 1991; Trowbridge & Mintzes, 1988), and 12-year-olds cite the head, appendages, and body covering as critical attributes when identifying vertebrates (Ryman, 1974).

As previously stated, there are two voices that act within the zoo. The personal context brought by the Visitor Voice is based on their nonspecialist naming of animals, while the personal context of the Zoo Voice is founded in a background of zoological and taxonomic education. For example, the zoo names the specimens at the lowest category in terms of the zoological taxonomy of the species and provides the scientific name (in Latin) and the formal common name. The visitors provide a vernacular name for the specimen and, unless the animal is a particularly well-known specimen like the Koala, do not use the species name. Visitors do recognize that animals belong to groups such as *bear* or *bird*. Therefore, the debate on appropriate taxonomic theory and nomenclature of the animals presented at the zoo may not be relevant or meaningful to the majority of zoo visitors. Zoological nomenclature encapsulates taxonomy which is concerned with component properties of the different types of animals which are *natural categories (monophyletic groups)* or *kinds*. Members of a *natural kind* share essential properties derived from a common ancestry with other members (Lakoff, 1987). Taxonomic judgments cannot be made without an understanding of the anatomy, physiology, genetics, and behavior of organisms. In fact, zoological taxonomy, including the ability to recognize the form and function of features and their morphological similarities, is a component of the science curriculums. Zoos play a key role in teaching visitors animal form and function and morphology. Therefore, zoos must assist visitors in understanding and recognizing simple morphological and taxonomic terms and identifying animals based on morphological attributes and similarities.

Identifying Animals

Visitors enter an animal collection already knowing something about the specimens and bring a unique level of relevant knowledge and interest (Linn, 1981). Therefore, when visitors enter an exhibit, they begin to view and make sense of the specimens using some preexisting knowledge of at least "everyday animals." The zoo visitor attempts to interpret animals in the light of their personal experiences with domestic animals at home and in their childhood (Cheek & Brennan, 1976). Visitors may prefer the zoo animals that evoke memories, which can provide a focus for dialogue at exhibits. The exhibit dialogue is likely to be formed from prior knowledge and be a response to the animals. The visitors' reaction to the animal in the exhibit is different from the reaction they would have to the same animal if they encountered the animal in a natural setting. The animal on display is *safe* and there to be observed.

Even though we do not know the level of zoological taxonomy that visitors spontaneously apply to an animal collection, previous studies do provide some information about the names used by visitors. Visitors refer to the animals that they see, but what names do they use? The zoo is a place where additional names for animals are acquired even if further cognitive development does not occur. Ethnobiologists have established that people use genus term to name organisms in their local environment. However, research has not determined the names visitors use when they are confronted with known and unknown animals during a zoo visit. The names visitors use may indicate their level of knowledge (Clayton, Fraser, & Saunders, 2009) and their topics of conversation. Visitor groups use genus terms, for

Table 6.1 The familiar names used by a family during a visit to the London Zoo grouped according to zoological taxonomy

Class	Order	Family	Genus	Species
Birds	Grub[b]	Monkey	Crane	Red-crowned crane
Worm	Echidna	New-world monkey	Rhinoceros	Asian lion
Fish	Bat	Lizard	Tiger	Giant woodpecker
Sea anemone	Grasshopper	Fly	Lion	Lord Derby's woolly opossum
Starfish	Cricket	Snake	Penguin	Zonure
Coral[a]	Stick insect	Cat	Woodpecker	Blue-tongued skink
Spider	Cockroach	Locust	Mynah bird	Indian cobra
Millipede	Seal	Tarantula	Viper	Splotched genet
Sea urchin	Beetle	Hermit crab	Cobra	Golden lion
Jellyfish			Skunk	Tamarin
			Chipmunk	Bird-eating spider
			Chimpanzee	
			Eel	
			Pike	
			Catfish	
			Batfish	
			Cobra	
			Plaice	
			Piranha	
			Giraffe	
			Zebra	
Total: 10	Total: 9	Total: 9	Total: 21	Total: 11

[a]Subclass
[b]Stage of development

example, rhinoceros, cobra, and giraffe, most often to name animals. Children and visitor groups with children use common or everyday names such as "monkey" and class terms such as "bird," "fish," and "insect" as basic terms for animals. Therefore, children and visitor groups are using basic, everyday, common names in terms of their taxonomic understanding (Cameron, 1994). Visitor groups do not use species names very frequently during their conversations, but species names are the ones most often provided by the zoos.

During a family visit to the London Zoo, three brothers, a grandmother, an uncle, and an aunt's conversations were recorded. The family used genus names more often than any other taxonomic group name (Table 6.1), but they did not name class, order, family, and species as often. Table 6.1 is a representation of the names this family group assigned to the animals during the visit. These may not be typical of every family group or of every group that visits the zoo, but they do provide a look at some naming techniques employed by the group. If these are the everyday and vernacular terms used by zoo visitors, there is a mismatch between visitors' terms and the terminology provided by the zoo's label. What might be the differences in conversational content about animals if the zoo's labels used the terminology applied by visitors? Would visitors use the labeling more extensively? Fourteen percent of the conversations within the family were based on terminology gleaned from labels. The 14% was when the family obtained a name or confirmed the animal's identification. We can only speculate whether the visitors would have used labels more and discussed their content, topics like conservations for instance, had the language been that of their experience, not that of technical zoology. We are not saying zoos should not provide taxonomically correct scientific names, but we are suggesting that zoos include common terminology. Work on classification of animals should be based on the naming system used by the children and that of zoology developed from it. Moreover, when designing labels,

> Exhibit designers need to pay attention to not only what is said, but how. For example, are the parents redirecting 'off task' comments? Are children's ideas responded to and incorporated into the conversation? Do all of the visitors in the group have a role in the conversation? What kinds of actions are shared with others and mimicked? How does the label change the kinds of activities, actions, and comments that are allowed? The answers to these questions help point to the ways in which visitors are framing the exhibit and are an important evaluation tool that helps to move evaluation beyond the necessary questions of navigation and understanding toward deeper conceptions of the museum and its role in informal education. (Atkins, Velez, Goudy, & Dunbar, 2009)

Young children use only one name for one animal and fail to recognize the exemplars that organisms have that would include the organism within the same group. This is a transitivity issue. For example, in the pike exhibit, Michael does not realize that the pike and eel are fish, but he does recognize that they have similar exemplars as fish. However, his grandma fails to explain to Michael that the pike and eel are fish:

Pike

Michael:	Look at that, Grandma!
Grandma:	What is it?
Michael:	I don't know.
Grandma:	What does it look like?
Michael:	A fish.
Aunt:	It's a pike! There's the eel! Look! Michael! The eel's half buried at the bottom!

When Michael is asked to identify a catfish as a fish, he is not able to make a connection because of the catfish's exemplars. The exemplars do not match his current beliefs about fish. This may be due to his uncle's suggestion that these fish meow because their name is catfish:

Catfish

Uncle:	See these in there? They are cat fish, can't you hear them meowing?
Michael:	I don't like these fish.
Aunt:	Aren't these fish?
Michael:	No! They are more like different fish. These sort of fish are not interesting.

However, in the tarantula exhibit, Michael does recognize the exemplars of a spider and is therefore able to identify the tarantula as a spider:

Tarantula

Michael: What's this one, Aunt Susan? Hey! There is a spider! Hey Grandma! Look at that!

Grandma: Ergh!

Michael: How did London Zoo manage to catch that?

When exhibits are put together in close proximity and the organisms within the exhibits look physically similar, children often believe that the animals must be in the same group. For example, in the conversation below, the sea anemones are displayed near the jellyfish. Michael shows that he is able to think in an abstract manner by identifying the anemone as an upside down jellyfish. Because of transitivity, Michael is able to recognize that the anemone and jellyfish are related but does not have the cognitive skills to separate the concepts of sea anemone and jellyfish. The aunt asks about the jellyfish next to the sea anemone exhibit to get Michael to think about the relationship. However, she tells Michael that the sea anemone does eat fish, that is, only eats small bits in the water. This may be true in the exhibit but is not indicative of the species. Sea anemones eat fish, shrimp, isopods, and plankton:

Sea Anemones

Aunt: What about the red things on the rocks?

Michael: What are they?

Aunt: Sea anemones, can you see the mouth? There, in the middle of the tentacles

Michael: What do they eat then, fish?

Aunt: No, they eat small bits in the water. Have you heard of jelly fish?

Michael: Yes, so when the sea anemone turns over it's a jelly fish?

Examples of conversations can be used to illustrate the way in which visitors name animals. Michael's family identifies and categorizes animals using prior knowledge and experiences. If an individual in the group fails to provide an answer, then the individual turns to another member of the group for an answer or another individual may interrupt the conversation with further information:

Tarantulas

Michael: What's this one, Aunt Susan?

Aunt: I am not sure.

Michael: Grandma what is it?

Grandma: The sign says it's Lord Derby's woolly possum.

Bird-Eating Spider

Michael: I wonder what's in there?

Mum: It's a huge big spider.

Aunt: It's a bird eating spider the label says.

Identifying animals in collections requires matching the specimen with one known name. Matching an unknown animal with a known name may be difficult for children. The extent to which children are able to express opinions about the classification of exhibited animals and justify their choice is largely unknown. During experimental laboratory conditions, children from 6 to 11 years old are unable to abstract critical attributes and use a wide range of organizational strategies to categorize (Bruner, Olver, & Greenfield, 1966). Some concepts are *fuzzy* or unclear. *Fuzzy sets* are concepts that are difficult to define because the defining boundaries between the concepts are *fuzzy* or not distinct (McCloskey & Glucksberg,1978). The person attempting to categorize asks themselves whether the object is a representative of a category or not. In a laboratory test situation, young children base their inferences about animal attributes on category membership and not just on names (Markman, 1989).

An example of a *fuzzy set* is a dolphin. Children have to resolve whether the animal is a fish or mammal. Dolphins look and act like a fish and they live in water, but they are not fish. Dolphins are members of the category mammal. Children must use deductive reasoning to define a dolphin as a mammal. Children are told that a dolphin *is* a member of the category *mammal*. Therefore, children deduct that a dolphin must have the characteristics of *mammals*, such as breathe with lungs, bear live young, and produce milk for their young. These characteristics are in contrast to the perceptual information gleaned from body shape and habitat, which resemble the category fish. When people see animals, they may not always look for the correct categorical features. Instead, they may adopt a nonanalytic approach that leads them to look for overall similarities between known concepts and the unknown object (Markman, 1989). When children identify animals based on overall similarities instead of specific exemplars, the children might be prevented from "prematurely settling on dimensions that might turn out to be wrong" (Markman, 1989; Wattenmaker, Nakamura, & Medin, 1988). For example, when Michael sees a bateleur eagle during the zoo visit, he names it "bird" but does not categorize it any further. Presumably, he does not use the name bateleur eagle because the specimen did not match any specific bird exemplars that he held.

The conversations that take place among children and their accompanying adults frequently consist of naming individual animal specimens (Hage, 1993; Hensel, 1987; Rosenfeld, 1980; Taylor, 1986). However, when a child asks for the name of an animal during a zoo visit, the child may not be categorizing the animal, they may only be acquiring information about its dimensions (Donaldson, 1978). For example, Michael's recognition (below) of the animal which he identified as a bird is a precursor of understanding categorization:

Bateleur eagle
Michael: It's sad!
Mum: Why is it sad?
Michael: He's got no other birds.

Animal Behavior and Anatomy

Research has documented that during a zoo visit, children and adults spontaneously comment about the animals' characteristics, behavior, etc. (Tunnicliffe,
1995). Visitors' spontaneous comments reveal that children notice particular animal
characteristics, and their comments are based on knowledge of themselves (Carey,
1985; Yorek, Sahin, & Aydin, 2009). This data may be a basis for zoos to use in their
interpretation and educational programs but should also be used to develop visitors'
knowledge into a deeper understanding of animals' form and function. Zoos should
begin using terms with which the visitors are familiar (monkey, bird, fish, etc.) in
addition to taxonomic terms.

When children visit the zoo, they need to know the names of the anatomical
parts of animals so that they may abstract out the criterial attributes of the animal's
taxonomic nomenclature. Being able to identify the parts of the whole organism is a
transitivity problem. Children are more likely to group animals taxonomically when
the animals have obvious external anatomy in common (Tversky, 1989). Young
children are quicker to notice the differences in animals when the organisms possess
prominent external anatomy, such as differing shapes or large size. Therefore,
children are likely to recognize and categorize animals based on very pronounced
shape, size, and external anatomy such as a large tail, the trunk of an elephant, or
the wing of a bird. For example, Michael (below) identifies an animal as a *shark*
by pinpointing several salient features that to him define a *shark*. However, when
confronted with identifying another organism, that is, in the same tank, as a shark,
Michael is able to discern that the batfish is not a shark because he declares that it
does not possess the requisite attributes:

Michael: Yes, these are sharks, look!
Aunt: How can you tell?
Michael: Because they are dead [sic] long and have tails like that. That's not a
           shark though (bat fish). I expect that's its food.
Aunt: Why is that not a shark?
Michael: That's their food probably.
Aunt: But how do you know it's not a shark?
Michael: Because it's flat.

Visitors interpret the animals' behavior and anatomy through prior knowledge
and personal context and use human anatomy and behavior as a reference point.
This phenomenon is anthropomorphism, and during conversations, visitors express
their anthropomorphic thoughts as affective comments, such as "He looks like your
uncle." or "That is how you act sometimes." The popularity of certain zoo animals
such as the giant panda may be due to the panda's anthropomorphic characteristics.
Young children employ anthropomorphic terms in their explanations of the behavior
and anatomy of animals, but by the age of 10, children adopt a biological perspective
regarding animals and recognize that animals posses certain attributes (Carey, 1985).

Attitudes, Emotional Connections, and Culture

Because "Zoos are a metaphor for our attitudes to and relationships with Nature" "(Hancocks, 2001), exhibit designers must consider visitors' attitudes and relationships with nature. If zoo exhibit designers make assumptions about visitors' attitudes to animals, the assumptions may color their approach to designing the exhibits. Not all children have a positive approach to animals and the environment. Some children are afraid of animals. The fear of certain animals is innate (Bennett-Levy & Marteau, 1984) and may be based on a self-perceived danger of the animal (Gray, 1971) or may be imposed by a parent's fears. Children may be influenced by the attitudes of their accompanying adults who are more inclined to hold *negativistic, doministic,* and *naturalistic* attitudes toward the animals. A *negativistic* attitude is when a person shows an active avoidance of animals due to indifference, dislike, or fear (Kellert & Berry, 1980). Kellert and Berry found that school children and children who had visited a zoo recently scored the highest in *negativistic* feelings toward animals. Thus, suggesting that visiting a zoo and formal learning about animals have very little positive influence on children. A *doministic* attitude is when people are interested in mastery and control of the animals, such as hunting and sporting situations. Children's *negativistic* and *doministic* feelings decrease as children get older (Westervelt, 1984). Children's feelings of negativity, fear, and domination may be a function of development and not of external influences like a zoo visit. Simply, zoo visits provide children with an outlet to express their feelings. Visitors with *naturalistic* attitudes are interested in the nurturing of the environment and learning about other species. Therefore, the conversations of children without an accompanying adult are likely to be different in content than those of children viewing animals with an adult.

A fear of animals peaks at about 4 years of age (Seligman, 1971). Adults seem to be fascinated with reptiles but find them to be the least attractive animal group (Bitgood, Benefield, Patterson, & Nabors, 1986). Fifth graders (10- to 11-year-olds), who are pet owners or whose classrooms contain live animals, have more positive attitudes to animals and to the environment than other students (ten Brink, 1984). However, US children between the ages of 6 and 10 years regard animals as "little friends" (Kellert, 1985, p. 30), and 6- to 10-year-olds were the most "... exploitive, unfeeling, and uniformed of all children in their attitudes toward animals" (Kellert, p. 31). A survey completed in the USA found that zoo enthusiasts were only marginally more oriented toward wildlife than nonzoo visitors but did have a stronger affection for individual animals (Kellert, 1980). This feeling, Kellert argued, indicated that zoo visitors were motivated by a general fondness for wildlife and not an innate curiosity about or attraction to wildlife or nature. Instead, survey participants were more concerned about issues of animal rights and welfare.

Visitors' preexisting attitudes toward animals affect where they focus their attention. The attributes of the animals determine where the visitor looks and talks about. Visitors focus on one attribute with which they are familiar or have a preconceived opinion or feeling. For example, a visitor may perceive an animal as ugly or dangerous (Bitgood et al., 1986; Bitgood, 1992). However, seeing "pretty animals" is one of the four reasons people cite for visiting a zoo (Kellert). The visitors' preexisting mind-set about the animals affects their interpretation of the exhibits (Whittall, 1992). These attitudes act as a perceptual filter (McManus, 1989; Wittlin, 1971) and create an emotional barrier between the visitor's general observations of the animals and the attributes upon which the visitors focus. Visitors are motivated during the zoo visit by their interest in animals, empathy and idealism of animals, and their belief in the natural state of animals (Hills, 1995). Furthermore, the emotional attitude related to the horror and concern of animal rights at the expense of the human species (Carruthers, 1992) may affect the manner in which visitors interpret the animals and the exhibits.

Other key factors that affect visitors' attitudes toward animals are the animals' behavior, the aesthetics of the exhibit, and the visitors' self-image. Humans are intrinsically fascinated with other animals and have a desire for human–other animal contact. This affective factor in live animal exhibits affects the visitor's responses to the exhibit (Finlay, 1986) and the resulting conversations. Animals that are kept in enclosures evoke emotional responses because of the ethical dilemma of capturing and enclosing animals for entertainment (Bostock, 1993). Visitors prefer naturalistic settings that do not have visible signs of artificial constraint such as bars and cages. Therefore, the way in which institutions present their exhibits may affect what the visitors notice, how they feel about the specimen, and what recollections they have of the visit. Moreover, visitors bring their own self-images which may influence the emotive reaction the visitors express (Morris & Morris, 1966).

Taking into account visitors' prior knowledge is important as zoos define their voice and develop visitors' knowledge. Zoos must remember that visitors bring knowledge that is "outside of the direct context and motivations surrounding the zoo visit itself" (Wagoner & Jensen, 2010, p. 73). Therefore, zoo educators should be aware that

> The cultivation of pre-visit representations of animals, habitats and the environment occurs over an extended period of time through the influence of multiple sources, including formal education and mass media. Education within the zoo must interact with such pre-existing ideas in the process of visitors' development of a new understanding of animals and their environments. (Wagoner & Jensen, 2010, pp. 73–74)

Kellert (1985) suggests that zoo educators enhance children's knowledge by focusing on the affective domain and emphasizing emotional concern and sympathy for animals.

The Visitor Voice within zoo exhibits should also be of concern to teachers. During a study completed by Patrick, Matthews, and Tunnicliffe (2011), preservice teachers listened to 636 students' conversations as the students participated in a field trip. The preservice teachers determined that school groups' conversations were focused on management, naming organisms, and locating organisms. Children in school groups were less interested in information about the animals and made few comments about conservation. After listening to the children's conversations, the preservice teachers thought that learning did not occur, the field trip was a waste of time, students were not interested in the visit, and chaperones did not do anything. Based on these observations, we can conclude that understanding the Visitor Voice is important when planning field trips. Therefore, Chap. 7 is a review of the Visitor Voice.

References

Anglin, J. M. (1977). *Word, object, and conceptual development*. New York: W.W. Norton.

Atkins, L. J., Velez, L., Goudy, D., & Dunbar, K. N. (2009). The unintended effects of interactive objects and labels in the science museum. *Science Education, 93*, 161–184.

Atran, S., & Medin, D. L. (2008). *The native mind and the cultural construction of nature*. Boston: MIT Press.

Bell, B. F. (1981). When is an animal, not an animal? *Journal of Biological Education, 15*, 213–218.

Bennett-Levy, J., & Marteau, J. (1984). Fear of animals: What is prepared? *British Journal of Psychology, 75*(1), 37–42.

Bitgood, S. (1992). The anatomy of an exhibit. *Visitor Behaviour, 7*(4), 4–14.

Bitgood, S., Benefield, A., Patterson, D., & Nabors, A. (1986). *Understanding your visitors: Ten factors that influence visitor behaviour*. Jacksonville, AL: Psychology Institute, Jacksonville State University.

Bostock, S. C. (1993). *Zoos and animal rights*. London: Routledge Kegan Paul.

Braund, M. (1991). Children's ideas in classifying animals. *Journal of Biological Education, 25*(2), 103–109.

Brown, R. (1958). How shall a thing be called? *Psychological Review, 65*(1), 14–21.

Bruner, J. (1983). *Children's talk. Learning to use language*. Oxford, UK: Oxford University Press.

Bruner, J. S., Goodnow, J. J., & Austin, G. A. (1956). *A study of thinking*. New York: Wiley, Science Editions, Inc.

Bruner, J. S., Olver, R. R., & Greenfield, P. M. (1966). *Studies in cognitive growth*. New York: Wiley.

Cameron, L. (1994). Organising the World: children's concepts and categories, and implications for the teaching of English. *ELT Journal, 48*(1), 28–39.

Carey, S. (1985). *Conceptual change in childhood*. Cambridge, MA: MIT Press/Bradford Books.

Carruthers, P. (1992). *The animals issue: Morality in practice*. Cambridge, MA: Cambridge University Press.

Cheek, N., Jr., & Brennan, T. J. (1976, September). *Some social-psychological aspects of going to the zoo implication for educational programming*. In Proceedings of the American Association of Zoological Parks and Aquariums annual conference, Wheeling, VA.

Child, D. (1985). *Psychology and the teacher* (4th ed.). London: Cassell.

Clayton, S., Fraser, J., & Saunders, C. D. (2009). Zoo experiences: Conversations, connections, and concern for animals. *Zoo Biology, 28*, 377–397.

Crisci, J., McInerney, I., & McWethy, P. (1994). *Order and diversity in the living world: Teaching taxonomy and systematics in Schools*. Reston, VA: National Association of Biology Teachers.

Donaldson, M. (1978). *Children's minds*. Glasgow, UK: William Collins Sons and Co. Ltd.

Driver, R., Squires, A., Rushworth, P., & Wood-Robinson, V. (1994). *Making sense of secondary science: Research into children's ideas*. London: Falmer Press.

Finlay, T. (1986). *The influence of zoo environments on perceptions of animals*. Unpublished Master's thesis, Georgia Institute of Technology, Atlanta, GA.

Gibson, J. (1979). *The ecological approach to visual perception*. London: Houghton Mifflin.

Gray, J. A. (1971). *The psychology of fear and stress*. London: Weidenfield and Nicholson.

Hage, S. (1993). Kids 'talk' to the animals. Family visitor study at the zoo. *Journal of the International Association of Zoo Educators, 29*, 30–34.

Hancocks, D. (2001). Is there a place in the world for zoos? In D. J. Salem & A. N. Rowan (Eds.), *The state of the animals 2001*. Washington, DC: Humane Society Press.

Hensel, K. (1987). *Families in museums: Interactions and conversations at displays*. Unpublished Ph.D. thesis, Columbia University Teachers College, New York.

Hills, A. M. (1995). Empathy and belief in the mental experience of animals. Reviews and research reports. *Anthrozoos, 8*, 132–142.

Inhelder, B., & Piaget, J. (1964). *The early growth of logic in the child: Classification and seriation*. London: Routledge & Kegan Paul.

Keil, F. C. (1979). *Semantic and conceptual development. An ontological perspective*. London: Harvard University Press.

Kellert, S. R. (1980). *Activities of the American public relating to animals*. Washington, DC: US Government Printing Office.

Kellert, S. R. (1985). Attitudes towards animals: Age-related development among children. *Journal of Environmental Education, 3*, 29–39.

Kellert, S., & Berry, J. K. (1980). *Knowledge, affection and basic attitudes toward animals in American society, Phase III*. Washington, DC: Fish and Wildlife Service.

Lakoff, G. (1987). *Women, fire and dangerous things*. Chicago: University of Chicago Press.

Linn, M. (1981). *Evaluation in the museum setting. Focus on expectations*. Washington, DC: National Science Foundation.

Lowell, W. (1979). A Study of hierarchical classification in concrete and abstract thought. *Journal of Research in Science Teaching, 16*(3), 255–262.

Lucas, A. M., Linke, R. D., & Sedgwick, P. P. (1979). School children's criteria for 'alive'. A content analysis approach. *Journal of Psychology, 103*, 103–111.

Macnamara, J. (1982). *Names for things: A study of human learning*. Cambridge, MA: MIT Press.

Markman, E. (1989). *Categorization and naming in children: Problems of induction*. Cambridge, MA: MIT Press.

Markman, E., & Seibert, J. (1976). Classes and collections: Internal organization and resulting holistic properties. *Cognitive Psychology, 8*(4), 561–577.

McCloskey, M. E., & Glucksberg, S. (1978). Natural categories: Well defined or fuzzy sets? *Memory and Cognition, 6*(4), 462–472.

McManus, P. M. (1989). What people say and how they think in a science museum. *Heritage Interprtation*. London: Belhaven Press.

Mercer, N., Dawes, L., Wegerif, R., & Sams, C. (2004). Reasoning as a scientist: Ways of helping children to use language to learn science. *British Educational Research Journal, 30*(3), 359–378.

Mintzes, J. J. (1989). The acquisition of biological knowledge during childhood: An alternative conception. *Journal of Research in Science Teaching, 26*(9), 823–824.

Mintzes, J. J., Trowbridge, J., Arnaudin, M., & Wandersee, J. (1991). Children's biology: Studies on conceptual development in the life sciences. In S. Glynn, R. Veaney, & B. Britton (Eds.), *Psychology in learning science* (pp. 179–201). Hillsdale, NJ: Lawrence Erlbaum.

Morris, D., & Morris, R. (1966). *Men and pandas*. London: Hutchinson.

Ninio, A. (1980). Picture-book reading in mother-infant dyads belonging to two subgroups in Israel. *Child Development, 51*, 587–590.

Ninio, A., & Bruner, J. (1978). The achievement and antecedents of labelling. *Journal of Child Language, 5*, 1–15.

Ollerenshaw, R., & Ritchie, R. (1998). *Primary science: Making it work*. London: David Fulton.

Patrick, P., Matthews, C., & Tunnicliffe, S.D. (2011). Using a field trip inventory to determine if listening to elementary school students' conversations, while on a zoo field trip, enhances preservice teachers' abilities to plan zoo field trips. *International Journal of Science Education*. Available at: http://www.tandfonline.com/doi/abs/10.1080/09500693.2011.620035, 1–25, iFirst article.

Piaget, J., & Inhelder, B. (1969). *The psychology of the child*. London: Routledge and Kegan Paul.

Rinsland, H. D. (1946). *A basic vocabulary of elementary school children*. New York: Macmillan.

Rosch, E., & Mervis, C. B. (1975). Family resemblances: studies in the internal structures of categories. *Cognitive Psychology, 7*, 573–605.

Rosenfeld, S. (1980). *Informal learning in zoos: Naturalistic studies on family groups*. Unpublished Ph.D. thesis, University of California, Berkeley, CA.

Ryman, D. (1974). Children's understanding of the classification of living organisms. *Journal of Biological Education, 8*(4), 140–144.

Seligman, M. (1971). Phobias and preparedness. *Behaviour Therapy, 2*, 307–320.

Sherwood, K. P., Rallis, S. F., & Stone, J. (1989). Effects of live animals vs. preserved specimens on student learning. *Zoo Biology, 8*, 99–104.

Smiley, S. S., & Brown, A. L. (1979). Conceptual preference for thematic or taxonomic relations: A nonmonotonic age trend from preschool to old age. *Journal of Experimental Child Psychology, 28*, 249–257.

Smith, E. E., & Medin, D. L. (1981). *Categories and concepts* (Cognitive Science Series). London: Harvard University Press.

Tamir, P., & Sever, A. (1980). Students' attitudes toward the use of animals in biology teaching. *The American Biology Teacher, 42*, 100–103.

Taylor, S. M. (1986). *Understanding processes of informal education: A naturalistic study of visitors to a public aquarium*. Unpublished Ph.D. thesis, University of California, Berkeley, CA.

ten Brink, B. L. (1984). *Fifth grade students' attitudes toward ecological and humane issues involving animals*. Unpublished Ph.D. thesis, University of Texas at Austin, Austin, TX.

Trowbridge, J., & Mintzes, J. J. (1988). Alternative conceptions in animal classification: A cross age study. *Journal of Research in Science Teaching, 26*(7), 547–571.

Tunnicliffe, S. D. (1995). The content of conversations about the body parts and behaviours of animals during elementary school visits to a zoo and implications for teachers organising field trips. *Journal of Elementary Science Education, 7*(1), 29–46.

Tunnicliffe, S. D., Gatt, S., Agius, C., & Pizzuto, S. A. (2008). Animals in the lives of young Maltese Children. *Eurasia Journal of Mathematics, Science & Technology Education, 4*(3), 215–221.

Tversky, B. (1989). Parts, partonomies, and taxonomies. *Developmental Psychology, 25*(6), 983–995.

Vygotsky, L. S. (1962). *Thought and language*. Cambridge, MA: MIT Press.

Wagoner, B., & Jensen, E. (2010). Science learning at the zoo: Evaluating children's developing understanding of animals and their habitats. *Psychology and Society, 3*(1), 65–76.

Wattenmaker, W. D., Nakamura, G. V., & Medin, D. L. (1988). Relationships between similarity-based and explanation-based categorization. In D. Hilton (Ed.), *Contemporary science and natural explanation: Commonsense conceptions of causality* (pp. 204–240). New York: University Press.

Westervelt, M. (1984, December). A provocative look at young people's perceptions of animals. *Children's Environment Quarterly, 1*(3), 4–7.

Whittall, R. (1992, September). *A walk on the wildside*. In Proceedings of the American Association of Zoological Parks and Aquariums annual conference, Toronto, Canada.

Wittlin, A. (1971). Hazards of communication by exhibit. *Curator, 14*(2), 138–150.

Yorek, N., Sahin, M., & Aydin, H. (2009). Are animals 'More Alive' than plants? Animistic-anthropocentric construction of life concept. *Eurasia Journal of Mathematics, Science & Technology Education, 5*(4), 369–378.

Chapter 7
Visitor Voice

At present, educational programmes do not make allowance for gender, age and social composition of the groups in which students work when within the museum. It is clear that the influence of the adult on the conversational content of the students is important and different groups need a different emphasis in preparation and interpretation. (Tunnicliffe, 2000, p. 273)

Zoos, museums, homes, schools, parks, or streets are places where various types of conversations occur. These conversations are usually social conversations that focus on the people involved and their surroundings. During a zoo visit, the zoo conversations that take place are social but also include learning conversations. The learning conversations that take place at the zoo consist of formal and informal learning (Lucas, McManus, & Thomas, 1986). The learning conversations focus on the exhibit (McManus, 1987) and are an attempt to make sense of what the visitors observe. Different visitors have different voices, and within a group, each visitor has a voice. The content of the conversations generated when looking at live animal exhibits will be different for family groups (leisure) and school groups (educational). Moreover, the leader of the group will influence the conversation. The presence of an adult in the group connotes that the conversational behavior of the children will be directed to aspects of the immediate environment (McManus, 1989). The adults accompanying the children are typically family members, family friends, school staff, teachers, or chaperones such as pupils' parents, and these adults will play a significant role in influencing what the children observe. Tough (1977) states that

The experiences which children have of adults using language with them must play an important part in influencing the kind of interpretation that children will make of their everyday experiences. If for example, the adult is talking about particular detail in the environment, the structure of plants the shape and color of the rainbow, the reflections in puddles, then the child's attention is being drawn to objects that he might not have noticed had no one spoken to him about them, or, if he had, might have remained at a level of interpretation that did not require conscious awareness of detail . . . (p. 35)

Constructing meaning is a social activity (Bruner, 1990) and is configured by the bidirectional conversations that occur between people. A part of the Visitor Voice is the conversations that occur during a zoo visit. Children do not visit zoos alone,

P.G. Patrick and S.D. Tunnicliffe, *Zoo Talk*, DOI 10.1007/978-94-007-4863-7_7,
© Springer Science+Business Media Dordrecht 2013

and therefore their conversations are directed by the adults within the group. As discussed in Chap. 6, children do visit the zoo with some animal knowledge, which they acquire from their earliest years through observations and what they are taught. Therefore, children will construct the meaning of the visit from prior knowledge and will construct their own personal understanding of the organisms (Bloom, Hastings, & Madaus, 1971; Gilbert, Osborne, & Frensham, 1982) and organisms' biodiversity. In fact, learning the names of animals is a key part in attaining knowledge about biodiversity (Chap. 6). During the visit, the adults point out the object and name it. As discussed in Chap. 5, the zoo's exhibit design can further extend the learning of the visitor by facilitating group interactions and conversations.

Meaningful observations that lead to learning are composed of conversations. An analysis of conversations' form and function reveals that conversations are part of an enjoyable experience and vary among their fundamental components. Conversations are bidirectional and are social because they occur between two or more people. The conversations within a group that visit the zoo emerge due to the group's social composition. For example, the content of the conversations among primary school groups (4–12 years of age) is different within each subgroup of teacher–pupil group, chaperone–pupil group, and pupil–pupil group. But conversations recorded at the London Zoo show that overall the conversations among school groups and family groups are similar (Tunnicliffe, 1995). Because conversations are a sociocultural activity, the culture of the group's members also impacts the conversational content and focus. When visitors enter the exhibit, their entry narratives will vary, especially in multiethnic communities. Teachers in multi-faith schools in England know that children from certain faiths do not look at owls, for example, or go near pigs. Therefore, zoos and educators must understand the social role of individuals during dialogues at animal exhibits. Zoos and educators must recognize the contribution of each Visitor Voice and scaffold learning to ask questions, probe the visitor's answers, and structure the dialogue with learning objectives.

Form, Function, and Categories of Conversations

There are two component characteristics of conversations: *form* and *function*. The *form* is constructed from individual words that build strings of words, called utterances. The *function* of the utterances varies, depending on the social settings in which they are used. However, one of the functions of conversation is to represent the thoughts and experiences of the discussants. Therefore, analyzing the content of zoo visitors' conversations will indicate the individual words used (form) and the topics of conversation (function). The process of conversing turns confusion into order, enabling the participants to construct a clearer picture of the world (Britton, 1970). Language plays a role in cognitive growth which means conversations, dialogue, and discourse are crucial for learning and developing understanding (Vygotsky, 1978). The following conversation is an example of meaning making and scaffolding new knowledge within prior knowledge:

Toucan

Boy: He's scuttled back. Hello!
Mother: He's a toucan. He's making a noise.
Boy: I didn't know toucans were still alive.
Mother: Oh! Yes!
Boy: Everyone says they're not alive don't they?
Mother: Dodos aren't. Toucans are.
Boy: Dodos?
Mother: Dodos aren't—Dead as a dodo.
Boy: If a zoo put a dodo in captivity they would be alive.
Mother: They'd be able to keep it alive.
Boy: And there'd be dodos in the world.
Mother: But there aren't any.
Boy: There aren't any more.

Halliday (1980) categorizes conversations based on the core function of the dialogue. Conversational exchanges that are based on observations are referred to as *experiential* or just noting facts. *Logical* exchanges occur when the viewer/speaker seeks to create a connection between what is seen and prior knowledge. However, Halliday's attempt to categorize conversations reflects the content as determined by the speaker but does not consider the receiver and their reaction.

Visitors' conversations provide commentary about visitors' interests, a look at visitors' prior knowledge and experience, and information about the focus of their attention within an exhibit (Falk & Dierking, 1992). Moreover, visitors' conversations reflect the visitor's role in a group and offer a glimpse at how visitors use conversation to express their needs, gain control of the conversation, and get the attention of other group members (Britton, 1970). Children learn by listening to conversations, interacting with the people who talk to them, and identifying what is emphasized during the conversation (Tough, 1977). Discerning the general nature of the conversations and their form and function within the zoo context is necessary to understanding and analyzing visitors' prior knowledge, appreciation for animals, and perceptions of biodiversity and conservation.

While language has generic features, language should be studied within the context of its use (Sinclair & Coulthard, 1975). Studying language in isolation can provide a framework that becomes a starting point in understanding the tool needed to analyze conversations. Because conversations are verbal bidirectional interactions and are determined by the individuals or institution, a tool is needed to establish the form and content of the conversations that take place during visits to zoos. An established view of conversations will aid classroom and zoo educators in developing successful educational tools. We must delineate the composition, content, and form of family and school group conversations at animal exhibits. By examining the conversations that take place at exhibits, we can conclude if the message of the exhibit is received by the visitors and reflected in the conversation's content.

Hymes (1972) has suggested that there are five main aspects of a conversation: (1) a shared language, (2) a setting, (3) a participant's expected outcome, (4) a form (e.g., a debate, a discussion, a presentation), and (5) a topic which is usually influenced by a conversant or the location (e.g., zoo or museum). However, Ellis and Beattie (1986) further analyze conversational content and regard communication as a cooperative interaction which can be broken down into four phases: (1) the transmitter originates and codes the communication, (2) the physical transmission, (3) the receiver who receives and decodes the transmission, and (4) the comprehension of the transmission so that the information can be used in a meaningful communication. Therefore, language is not only given or transmitted, it is received, and conversations are the exchange of the perceptions of the communicator with a recipient about a topic, which in this book are organisms. For successful communication to occur, each participant must understand the transmission of the other party and be able to effectively *decode* the communication (Ellis & Beattie). Decoding the communication in zoos is crucial for developing educational interpretation because institutions may not transmit their interpretation in a language that the receivers are able to decode. However, the communication between the zoo and the visitor is not the only important place where decoding occurs. Individuals within a group of visitors may not share the same code and may not understand the transmissions from each other. The function of conversations determines the form; therefore, an analysis of the form becomes a tool which enables an analysis of function (Britton, 1970).

Conversations usually begin with a summons/answer sequence with the respondent replying (Hensel, 1987; Schegloff & Sacks, 1973). The utterances within the bidirectional summons/answer exchange are known as *adjacency pairs*. Therefore, language possesses a structure that consists of dialogic phases, grammatical structure, and operational units. Conversations are a form of discourse or language, and within conversations, the use of language can be determined through discourse analysis (Gee, 2010). In this chapter, we look at discourse as a function or result of the interactions within a zoo exhibit, and we use discourse analysis to make sense of the visitors' interactions with that context. As a function, discourse is composed of three major forms: (1) statement (declarative), (2) question (interrogative), and (3) command (imperative) (Sinclair & Coulthard, 1975). As a sensemaking tool, discourse analysis provides a look at the meaning making that occurs during the conversation and provides linguistic clues as to the intent of the transmitter or speaker. Specifically, the discourse must be analyzed within the context in which the conversation is generated (Sinclair & Coulthard, 1975). Within the discourse that occurs in a conversation, there is a relationship between the grammatical structure, function, and social purpose (Britton, 1970).

A conversation may also be analyzed by categorizing the conversation's structural complexity, content, and pattern. For example, the concept of *Motherese* is referred to as child-directed speech (CDS) in which adults talking with very young children modify their talk using "baby" words and simple structures that they believe mimic the way in which very young children talk. There are three levels of labeling conversations that take place in zoo exhibits. Level 1 is heard between adults and babies/toddlers. The following example takes place between a mother and a 3-year-old child. The mother begins the dialogue with a question that draws attention to the specimen, then gives the organism a name, encourages her child to repeat the name, and provides praise.

Kookaburra
Mother: Do you know what that is? It's a Kookaburra. Can you say Kookaburra?
Child: Kookaburra.
Mother: Well done.

Level 2 is characterized by the way in which adults talk with preschool children. Preschool children have developed further than repeating names, naming, or labeling, but the adult may still use CDS. For example, in the following discourse at a Toucan exhibit, the mother says birdie, but the 5-year-old child uses the more advanced term bird.

Toucan
Mother: Look! A birdie!
Child: Yeah. This bird has big nose.
Mother: Pretty birdie!

Level 3 is *child's talk* in which the conversation is initiated by and closed by the child. This is known as an inverse triadic dialogue (discussed in the next section). Level 3 dialogue is the opposite of the dialogue in Level 1. The following is a comparison of Level 1 and Level 3 dialogues. In Level 1 dialogue, the mother starts and ends the conversation as the information giver. In the Level 3 dialogue, a 7-year-old boy and his father are discussing elephant feces after the child asks a question.

Iguana (Level 1)
Mother: Look at these guys. He has a yellow head.
Child: Where?
Mother: It's right here!

African bush elephant (Level 3)
Child: What's that? It looks like elephant pooh?
Father: That's right. And there are beetles, which live in it.
Child: Ergh!

As exemplified above, the pattern of conversations alters with the child's age. Very young children are being talked to through CDS, and older children are taught labels. During a field trip, all three levels of content may be observed. Helping the children learn about the animals is carried out by family members or adults in different styles based on child's age. However, infrequently, the child's comments are used to reinforce the child's naming of an organism or its observations or as an opportunity for incidental teaching. Instead, primary school-aged children are provided with pertinent information in a declarative manner. Occasionally, teaching discourse takes place in the form of key questions from the adults and responses from the child with reinforcing praise from the adult.

The content of conversations that takes place in exhibits may also be categorized into four distinct categories (Tunnicliffe, 1993): (1) Exhibit access is making sense of the exhibit and finding something to observe. (2) Exhibit focus is observing the organisms' structure and behavior and seeking to categorize the observations. (3) Management focus includes organizing the group and directing their movement. (4) Social interactions are when the visitors respond to other individuals in the group (e.g., Yes, Mother, Mom, Miss). This category consists of irrelevant social conversations that allow the conversations in the first three categories to flow. Social conversations are not content dependent, but exhibit access conversations are content dependent and focus the visitors' attention on the exhibit. Exhibit focused conversations include comments on the structure of the exhibit. Management focused conversations are not associated specifically with the animals. The exhibit is a backdrop around which the speakers use discourse to manage social situations. Moreover, the types of conversations that occur within an exhibit are not mutually exclusive. They may have a dual role. An exhibit access comment may also be used to manage the group, or a social interaction may also be animal focused. For example, exhibit access conversations consist of words such as "look" which direct attention when the group enters the exhibit but also serve as a management focused directive to the group (Hensel, 1987). Furthermore, social interactions and management focused conversations may occur when the visitors are not physically in an animal exhibit but are between exhibits. The content of exhibit focused conversations may be determined by a direct observation of the animal within the exhibit or by indirect observations, which are the visitors' memories.

Discourse in the Exhibit

Science educators are aware of the importance of the social aspects of learning. Therefore, studies of zoo visitors have focused on interactions among families, while fewer studies have looked at discourse that takes place during field trips. During a zoo visit, families, groups with children, and school groups engage in adult-directed discourse. The content of the discourse ranges from declarative statements to *teaching talk* (Parker, 2007). For visitors using exhibit text, there is often an entry point that is identified by a declarative pronouncement. For example, a visitor may announce the animal's name and may state information, which they have read from the exhibit's labels. This type of label information parroting is referred to as Text Echoing (McManus, 1989). Visitors may use the exhibit's labels (see Chap. 5) as a cue to begin discourse, as a way of framing or guiding their interaction with the exhibit, or as an information resource that they frame within their own context. Zoo's labeling presents information about biological conservation science and the zoo's biological research and biological conservation research. If a zoo explicitly makes a connection between its exhibits and science action, the zoo will provide visitors with opportunities to encounter science learning and will encourage a broader view of the nature of science. The discourse that occurs as a result of the encounters that take place at the zoo is an important aspect of the public understanding of science and should be used to increase the science literacy.

The conversational discourse that occurs within the exhibits has exchange boundaries and teaching boundaries. Exchange boundaries have two elements of structure: (1) focus and (2) frame; and two moves (or acts): (1) focusing and (2) framing. Teaching boundaries have three elements of structure: (1) initiation, (2) response, and (3) feedback; and three moves (or acts): (1) opening, (2) answering, and (3) follow-up (Sinclair & Coulthard, 1975), which are known as triadic dialogue (Lemke, 1990). The discourse at zoo exhibits is an inverse triadic dialogue in which the child initiates the conversation, the adult responds, and the child closes the exchange and includes a variety of sequential categories that demonstrate the development of observation and interest. The following conversation validates the inverse triadic dialogue:

Lion
Child: Where is it? I don't see it.
Mother: I see it over there. Can you see it?
Child: Yes. It is big.
Mother: It is a lion.
Child: It looks like a big cat. [The child ends this exchange and begins a different exchange.]

This pattern of discourse may not be unique to zoos but does reflect the way children lead the discussion and demonstrate that adults are not familiar with the subject matter and cannot *talk to teach*. Children initiate many conversations, but the adults do not direct the exchanges. The majority of children's comments are observational or are a commentary, with little effort by adults to help the children scaffold knowledge and *talk science*.

Discourse analysis reveals that conversations are used for social purposes, to manage people, to provide information, and/or to acquire information. The same is true of teachers and learners who visit the zoo. Teaching and learning discourse consists of descriptions and observations and seeking fundamental information, locating the animal, naming the animal, and stating information or facts. What makes a conversation one of science teaching rather than an everyday information exchange? To answer this question, we must consider the discourse that occurs during quality teaching and learning. Discourse at zoo exhibits is framed by a variety of categories within a sequence which allow the observer to observe and develop their interest. The most important aspect of the zoo visit is to aid children in learning about animals. Teachers and parents facilitate this learning with different conversational interactions called *learning access conversations. Learning access conversations* use specific forms of discourse that may be spoken separately or in conjunction. Therefore, one or more forms may be used within a learning access conversation. The specific forms of discourse that occur within a zoo exhibit are the following:

(1)   **Focusing** Questioning at a new exhibit that focuses on the exhibit, behaviors, and/or anatomy. For example, "What about these?" "It's got spiky fur, look."

"They move quickly, don't they?" "There it is over there!" "Where is it?" "I don't see it!"

(2)   **Informing** Fact telling that informs the group. For example, "They live in the desert where it is hot." "The sign says it is mostly closely related to an elephant." "It's a vegetarian, it only eats plants." "This is the largest land mammal." "It eats fish." "It only lives in the Central American rainforest."

(3)   **Developing** Questioning that aids the child in developing the child's own thoughts about organisms. For example, "Why does it have spots?" "Why do you think it has such big teeth?" "Why are its eyes in front instead of on the sides of the head?" "What do you think he uses that trunk for?" "What are those long claws for?" "Which fish is the biggest?"

(4)   **Assessing** Finding out what the child is thinking. For example, "How do you know?" "How do you know it's a monkey by looking at it?" "How can you tell that is a bird?"

(5)   **Interpretive** Assisting others in understanding, reasoning, and justifying comments or names. For example, a boy explains that the animals are yellow because the teacher stated that the animals live in the desert. In another group, once the teacher leaves, a girl asks her friends to help her understand the teacher's comment that there are old-world and new-world monkeys.

(6)   **Feedback** All learners need feedback, such as praise, reprimand, or suggestion. For example, "You don't remember this from class? You have a short memory." "Yes, that's right insects." "Yes, they look like dogs, but they are not dogs." "You know better than to act like that."

(7)   **Terminating** Bringing the period at a particular exhibit and its associated discourse to a close. Teachers are particularly good at terminating. For example, "Right, come along." "Stop that we have to go!" "We are leaving!" "We must move on. We do not have time for that." "Let's go see the polar bears."

These forms of discourse may take place during free looking (without an adult) or may be adult guided and should provide children with exposure to the exhibit and learning. Learning at the basic level can be defined as an increase in knowledge and understanding. During a zoo visit, the goal of education is to aid visitors in learning the scientific interpretation of the organism, for example, developing an understanding of the organism's taxonomy, physiology, ecology, and conservation.

Zoos, like museums (Spicer, 1994), have the difficulty of balancing exhibit accessibility, scholarly research, and public erudition. For example, zoos are working toward research, preservation and conservation of endangered species (di Castri & Younès, 1996), and educating visitors, yet an analysis of the discourse of groups visiting zoos in the USA and England shows that conservation is rarely a topic of conversation (Patrick, Matthews, & Tunniclife, 2011; Tunnicliffe, 1994). Therefore, if zoos are teaching about animals, then the exhibit design should have a focus that enhances authentic learning. Exhibits should not provide a fantasy using animals with distinctive anthropomorphic features that elicit affective comments but instead

Table 7.1 An example of using a modified SOLO taxonomy to determine the extent to which groups use cognitive discourse

Speaker	Dialogue	Score
Girl:	It's beautiful	1
Teacher:	Do you know what it is?	2
Girl:	No. Cow, cow	2
Boy:	Sheep, sheep	2
Teacher:	Some sort of antelope I think	2
Teacher:	See if you can see a picture of some sort of striped animal	3
Boy:	Bongo, bongo	3
Teacher:	Lisa, what's the difference between this one here and that one (actual specimen and photograph on label)?	4
Girl:	That one's ...	3
Teacher:	Look at the patterns	3
Girl:	They're the same pattern	3
Teacher:	What sort of pattern?	4
Girl:	Black and white	2
Teacher:	No. It's spotted	3
	Total utterances = 14	Total = 37
		Rating = 2.6

should align exhibits with the topics about which the visitors talk. Visitors notice and comment about particular animal characteristics. Based on prior research, we know what children observe by listening to their comments, listening to how the adults respond, and/or observing what children and adults point out within exhibits. An essential part of planning authentic learning tasks is identifying the general topics upon which visitors' will comment. If these topics are taken as the starting point, then the authentic learning tasks can be planned around the conversational topics. Visitor talk is prompted by (1) location of the animal within the enclosure, (2) feeding and eating, (3) movements, and (4) attention-attracting behavior such as urinating and copulating.

Structure of Observed Learning Outcomes (SOLO) taxonomy may be used to analyze written responses and focuses on definite task sets (Biggs & Collins, 1982) and may be applied to verbal responses. However, SOLO taxonomy is an analysis designed for *performance* of a particular task and not as an overall allocation of reaching a specific cognitive level. Therefore, SOLO taxonomy may be adapted as an analytical tool to identify the discourse overheard in zoos. During an analysis of verbal discourse, a listener can discover whether adults are questioning children or children are questioning adults and determine children's responses. Children in verbal discourse are not writing their responses, but the answers are similar to those given in a written task. Using SOLO taxonomy makes it possible to identify the learning nature of visitor talk (Table 7.1).

The utterances within a discourse can be categorized, labeled, and quantified. Within a conversation, the utterances can be categorized and the occurrence of each category summated. The categories relevant to zoo visits are (1) social,

(2) prescience, (3) science information (statement or question), and (4) cognitive thinking (statement or question). Table 7.1 is an example of using the modified taxonomy to categorize a family's conversation at a Bongo exhibit. Once the conversation is coded, the categorical numbers can be summed and divided by the number of speakers to determine the level at which the discourse achieved. The higher the average, the more likely the discourse was to include questions and comments that involved cognitive thinking strategies (Blooms Taxonomy). This method provides an approximate indication of the nature of the conversation in terms of the educational discourse. In this example provided in Table 7.1, the family scores a 2.6, which means some of the discourse does include specific science content but may not be cognitively engaging.

Conversational content has distinct characteristics which help visitors identify animals and focus their interaction. If the visiting group is on a field trip, the form, function, and content of the discourse is dependent on where the child is in learning biology. If the visiting group is a nonschool group, the form, function, and content of the discourse depends on the level of biological knowledge of the visitors within the group. However, whether or not visitors in any group retain the information that is shared during the discourse is not clear. Therefore, analyzing the conversational content of the discourse does provide an indication of the visitors' psychological involvement and cognitive interactions. The conversational content should increase in complexity from identifying to describing to interpreting and eventually applying information (Borun, Chambers, & Cleghorn, 1996). Identifying encompasses naming animals using everyday terms such as cat or lion but does not include scientific names (*Panthera leo persica*) or specific common names (Asian Lion). When visitors reach the describing level, they may make associations between the exhibit's content and characteristics with personal experiences. In the interpreting and applying level, visitor discourse includes multiple word exchanges with each other, and visitors may correct other group members. Additionally, visitors apply the label information to previous exhibit information (this seldom occurs) and prior knowledge. In families and occasionally school groups, an adult assumes the role of information source and provides declarative discourse in which they provide information (identifying) and seldom provide time for children to respond with feedback (interpreting and applying). The conversational content may not measure learning but does provide a glimpse at the quality of the discourse that takes place and the minds-on interactions.

Discourse in informal settings, such as museums and science centers, are different from the conversations about animals that occur during a zoo visit. In museums, parents initiate interactions by questioning their children, but in interactive science centers, the conversation is dominated by accompanying adults. At the zoo, discourse within groups is *incidental,* and the children may initiate the dialogue. *Incidental teaching* opportunities that are presented by children may be ignored by the accompanying adults. Moreover, adults or children may initiate the dialogue and share their expertise in zoology. Adults may occasionally initiate the conversation in a *school-talk manner*, while children initiate dialogue by asking the location of an animal. The dialogue then can be utilized by an adult to develop language. The resulting discourse that occurs between the children and adults is different from those in the classroom. The discourse that occurs in the zoo does not resemble the discourse that occurs during formal classroom teaching but is similar to the conversations that transpire between children when they are involved in discovery learning and discussion.

Using Grounded Theory to Analyze Conversations

Grounded theory (GT) is a research method in which the researcher collects data and then develops a data code that is used to analyze the data. Therefore, GT can be used to analyze conversations within exhibits to conceptualize the emerging patterns of discourse. Glaser (2011) states that GT can be used to

> get out of the data to the emergence of conceptualization . . . trust in emergence by starting to constantly comparatively analyze data the first night after field notes are collected. Look for interchangeable indicators in the data as it is collected, as it is gathered, while using the constant comparative method. Once the pattern, latent in the data, is found, and indicators are saturated, name the pattern and conceptualization begins. Soon more patterns emerge and memos start relating them and a multivariate GT starts emerging around a core category. The research has gotten out and off the descriptive level of gathered data to conceptualization. GT is about concepts, not description. (p. 3)

By recording and listening to conversations, broad concepts of conversational content become apparent, and verbal concepts are identified as specific types of comments. Some of the verbal concepts or comments that occur during discourse at the zoo are behavioral comments, naming comments, taxonomy comments, comments about the animal's characteristics, orientation comments, management comments, information giving comments, information seeking comments, social comments, and exhibit focused comments. The following conversation is an example of the concepts or comments heard in a zoo exhibit with the comments labeled. However, this is not meant to say that one conversation determines the concepts found in conversations. The following conversation is used only as a representation:

Bearded lizard
Boy 1:   There's one and there's one and there's one (management).
Boy 2:   See that buffalo (naming) skull over there. That's from America (infor-
           mation) and there's a light bulb, too (exhibit design)!
Teacher: Can you see the animals clearly (management)?
Boy 1:   No, they blend in (characteristic).
Boy 2:   What do you call that (information seeking)?
Boy 2:   Camouflage (information giving).

During discourse, visitors frequently mention the animals' behaviors and characteristics. The four main concepts of visitors' behavioral comments are (1) the animal's location, (2) movement, (3) feeding-related behavior, and (4) intermittent behaviors that attract attention such as urination, copulation, and parental care. The following conversations are examples of the four behavioral comments:

Tortoise (location)
Boy: Tortoise!
Girl: Where?
Girl: Up there!
Girl: I see it. Can you see that red thing. Up there. You can see that, uhm, red thing. That red thing behind the rock.
Boy: Where do you see a red thing?

Tasmanian devil (movement)
Girl: Aren't they cute! Look at them running around and around and around!

African bush elephants (feeding-related behavior)
Girl: Look! It's eating a branch.

Dik-dik (intermittent behaviors that attract attention)
Girl: What is it doing (one dik-dik is licking the genitalia/anus of another dik-dik)? It is licking the other one!
Parent: Come on!
Girl: Ewww!

Moreover, the comments about the characteristics of animals occur as four concepts: (1) front end of the body (eyes, head, nose, ears, and neck), (2) dimensions (shape, size, number, color, and covering), (3) unfamiliar attributes (horns, excretory and reproductive organs), and (4) appendages (legs, arms, wings, and tail). The characteristics most frequently mentioned are the dimensions and the front end of the animals. The following conversations are examples of the four animal characteristics comments;

Cotton topped tamarins (front end of the body)
Girl: They have a small nose and a small mouth.

Cotton topped tamarins (dimensions)
Boy: That is a baby that is a baby
Teacher: Why?
Boy: Because they are small.

Chimpanzees (unfamiliar attributes)
Girl: Look at its pink bottom!

Elephants (appendages)
Boy: It's putting its tongue . . . I mean its trunk in its mouth!

Another GT technique changes the content of the conversations into numbers. Once the conversation is captured digitally, the conversation is converted to a hard copy. In short conversations, a count of the main categories of the content can be identified through the reiterative process of reading and rereading the transcript (Tunnicliffe, 1996). Once the categories have been identified, a count of the number of times the category is used can be determined. Categories may include artifacts, behaviors, exhibit access, biofacts, animal knowledge giving, name of specimens, information giving (geographical, technical name of behaviors), information seeking (questions from learner), teaching questions (to learner), social, and others. The conversation below is an example of utilizing the categories. A quicker technique based on the systemic network analysis that yields numerical results quickly is a Tunnicliffe Conversation Observation Record (see Chap. 10). While listening to voices in the zoo, a quick "tick" box instrument can be used to gain information about the main categories made at various animal exhibits.

Bearded lizard

Boy 1:	There's one and there's one and there's one (exhibit access).
Boy 2:	See that (exhibit access) buffalo skull (name of specimens, biofacts) over there. That's from America (information giving) and there's a light bulb, too!
Teacher:	Can you see the animals clearly?
Boy 1:	No, they blend in (behavior).
Teacher:	What do you call that [teaching questions (to learner)]?
Boy 1:	Camouflage (animal knowledge giving).

However, the analysis of longer conversations requires the development of a systemic network. The application of a systemic network allows for a statistical analysis of the sequence of dialogue spoken within a particular animal exhibit. A conversational unit consists of the conversations that occur within one exhibit, and the discourse begins when the visitors enter the exhibit and end when the visitors exit the exhibit. Once the conversations are transcribed, the discourse's conversational content can be worked out by a read and reread technique. Once the broad conversational categories are established, the researchers can produce a systemic network (Bliss, Monk, & Ogborn, 1983). A systemic network is a means of grouping or categorizing things, in this case conversations, into a parsimonious representation of the data. The systemic network preserves the relationships between categories in such a way that comparisons can be made between groups. The network can be regarded as a set of boxes that are filled with the parts of conversational discourse (Fig. 7.1). On the left side of network are categories that are highly specific items, and at the far right end are the main descriptors with numbers that label the most specific level of table categorization. A bar ("[") indicates that an attribute may be either/or but not a member of both categories, and a bracket ("{") indicates one of a number of categories which the member of the category may have.

Using Fig. 7.1, the conversations are coded using the number of the relevant category which is written above the appropriate words on the transcript. Each category of comment has a number allocated to it. A mention of a live animal was a 20 (see transcript below where the animal is referred to as "one"). An anatomical comment is a body part. If it were about the head or part of that such as the nose, the comments were coded as 43, which subsumed more detailed anatomical comments.

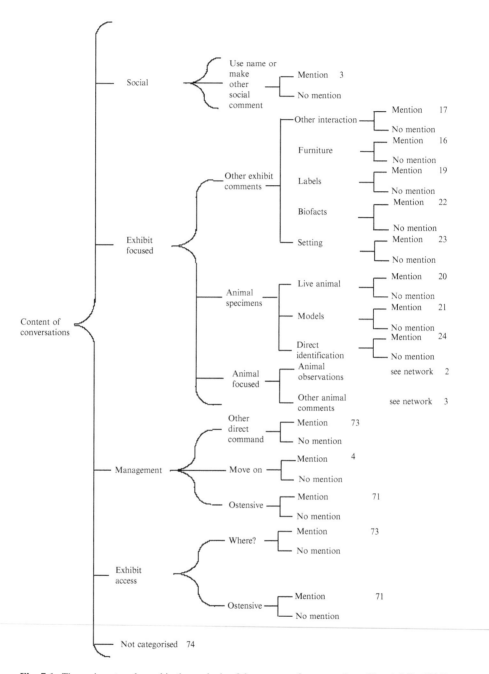

Fig. 7.1 The main network used in the analysis of the content of conversations (Tunnicliffe, 1996)

The numbers of the coding system are written over the relevant words. Not all the words have been categorized here for ease of reading. If more than one comment of a particular category (e.g., a name) occurred within a single conversation, it was not scored again. Hence, the analysis shows the number of conversations within which a topic is mentioned, not the number of overall times that a topic is mentioned. Some comments were categorized more than once. For example "look'" was categorized as a management statement as well as one of exhibit access because it was an ostensive remark. Although there are computer programs that can carry out such an analysis, the nuances, relationships, and the feel of the dialogue are lost. Each conversation unit was categorized with the appropriate number from the networks. Hence, the following conversation was coded in the following way:

Conversation at giraffes (there were several babies)

Girl	They are tall, aren't they?
Girl 2	Look at that giraffe!
Boy	That's as tall as a house!
Boy	They eat trees, they eat too much.
Boy 2	That tree used to have lots of leaves on it.
Boy 1	I'd only come past his knee!
Boy	How come they do not let the babies out?
Boy	Because they would easily go over.

The conversation is coded:
Coded conversation at giraffes (there were several babies)

Girl	21/50/13
	They are tall, aren't they?
Girl 2	72/55
	Look at that giraffe!
Boy	21/61
	That's as tall as a house!
Boy	21/36/12
	They eat trees, they eat too much.
Boy 2	18/12
	That tree used to have lots of leaves on it.
Boy 1	60/51
	I'd only come past his knee!
Boy	13/53
	How come they do not let the babies out?
Boy	12/23
	Because they would easily go over.

Applying this method to categorizing children's conversations within a zoo exhibit uncovers 74 categories in the network (Fig. 7.1). The four main super ordinate categories are *social comments*, *exhibit focused comments*, *management comments*, and *exhibit access* or *orientation comments*. A *not categorized* category is provided for uncoded topics such as security announcements. However, uncoded comments that refer directly to the exhibits are placed into *other exhibit* comments. For example, comments about other aspects of the exhibit (such as the rocks behind animal, a water feature, plants) are placed in the *other exhibit* category. Moreover, the systemic network uncovers four main categories of comments about body parts (Fig. 7.2): the front end of the body, the dimensions (shape, size, number, color, and covering), unfamiliar attributes such as horns and excretory and reproductive organ, and attributes which disrupted the body outline such as legs and a tail. The attributes of the body which were referred to most frequently were the dimensions (size, color, etc.) and the parts at the front end of the animals. The most frequently observed behaviors that were commented upon were the position of the animal in the enclosure, movement, feeding-related behavior and intermittent behaviors which attracted the attention of the visitors such as urination and parental care.

Children look at specific attributes of animals—parts of the body and behavior. There are four categories of body parts: (1) the front end (head and sense organs), (2) the dimensions of the animal (size, shape, color and body covering, and number of animals), (3) the disrupters to the outline such as tails and legs, and (4) unfamiliar organs such as horns and excretory and reproductive organs. Table 7.2 provides examples of discourse that yield these results.

The four main categories of behavior (Table 7.3) are (1) the location of the animal in the enclosure, (2) feeding-related behaviors, (3) movements (including locomotion but excluding specific movements named such as breathing), and (4) attention-attracting behavior such as urinating and copulating. The animal focused category was subcategorized into six subordinate groups: (1) interpretative comments, which included knowledge source comments such as questions and references to a source of the information proffered, human resemblances; (2) affective comments which included emotive responses such as "Ah!" or "Ugh" as well as comments about other attitudes, namely, human–animal interactions (and vice versa) and welfare comments; (3) environmental comments referring to the natural habitat or endangered status of the species; (4) comments about the animals' structure; (5) comments about the animals' behaviors; and (6) comments about the animals' names. Tables 7.4 and 7.5 provide examples of discourse that yield these results.

Tunnicliffe (1996) found that visitors give their opinion or ask a question in 53% of the individual conversations that take place in zoos (Table 7.6). Thirty-two percent of conversations involved anthropomorphic comments (It looks sad.), and 16% expressed concerns about the danger of the animal or what the visitor would like to do with the animal (Is it poisonous? I'd like to pet it.). Emotive attitudes such as "Ah! I like that! Gross! or Ugh!" occured in 32% of the conversations. Visitors named the organism or made a "naming" comment in 88% of the conversations, and 47% of the names were at the order/family level of zoological taxonomy. Twenty

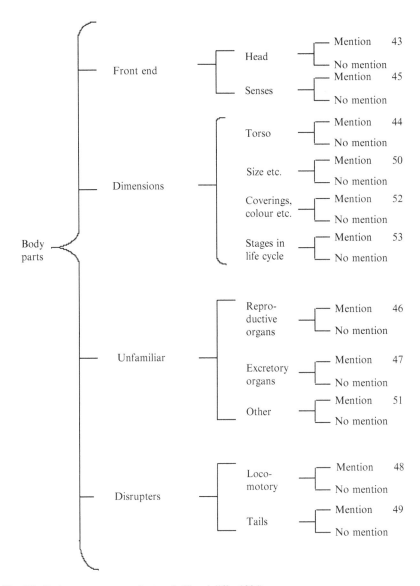

Fig. 7.2 Body parts segment of network (Tunnicliffe, 1996)

percent of the conversations refered to at least one body part. Sixty-six percent of the comments were on aspects of the exhibit. However, only 21% of the conversations made a direct reference to the exhibit's labels. Groups with adults had a higher occurrence of anthropomorphic comments than groups with children, except for comments about parts of the body (e.g., tails and legs that disrupted the outline of the body).

Table 7.2 Comments about body parts

Category of comments about body parts	Examples of conversations
The head and sense organs	*Cotton topped tamarins*
	Girl: They have a small nose and a small mouth
Body dimensions of the animal (e.g., size, shape, numbers)	*Cotton topped tamarins*
	Boy: That is a baby that is a baby
	Adult: Why?
	Boy: Because they are small
	Lizards
	Girl: Look at the size of it!
	Boy: It could be a small one
Unfamiliar and unusual (e.g., horns)	*Chimpanzees*
	Look at its pink bottom!
Disrupters to the outline (i.e., appendages or other parts of the body that stick out and often move, e.g., tail, legs)	*Elephants*
	It's putting its tongue. I mean its trunk in its mouth!

Table 7.3 The categories of conversation about behaviors

Category	Examples of conversation
Movement	*Zorillas in moonlight world*
	Girl: Aren't they cute! Look at them running round and round and around!
Position in the enclosure of the animal in relation to the exhibit	*Reptile house*
	Boy 1: Tortoise! Up there! I spotted it. You can see that red thing. Up there. You can see that, uhm, red thing, that red thing behind that rock. Boy 2: Where up there can you see that red thing?
Feeding	*Elephants*
	Girl: Look. It's eating a branch
Attractors	*Elephants*
	Girl 1: Look the elephant is doing a wee What a wee! It must have had a lot to drink What is the other one doing?
	Girl 2: Treading in wee! It's covering it up with sand!
	Giraffe (giraffe's head and neck is bent toward the floor)
	Girl 1: Does it have to be like that?
	Girl 2: Otherwise it will hit its head on the roof

Tables 7.3 through 7.6 show that during discourse the visitors not only comment about their observations of the animal, they interpret what they see by asking questions and making statements, explaining their observations in terms of previous experiences which are largely anthropomorphic in nature. Visitors express a range of affective comments and also exchange *management* comments often associated with control of the groups and social utterances, they most often using each other's name and acknowledging comments directed at them or calling the attention of an individual to a feature of the exhibit. Looking at other features of the exhibit is an important part of over half of the conversation. It is interesting that direct

Table 7.4 Main superordinate categories of the topics of conversation about body in a study of primary children visiting an English zoo (Tunnicliffe, 1996)

Category (formed from amalgamating data from individual categories of body parts)	Number ($n = 459$)	% of total conversations	% of conversations about behavior ($n = 280$)
Front end (head and sense organs)	77	17	28
Dimensions (body, size, life history, and coverings)	237	52	85
Unfamiliar parts (excretory and reproductive organs and other)	32	7	11
Disrupters to outline (tail and locomotory organs)	57	12	20

Table 7.5 Main superordinate categories about behavior in conversations (Tunnicliffe, 1996)

Category (formed by amalgamating data from individual categories)	Number ($n = 459$)	% of total conversations	% of conversations about behavior ($n = 301$)
Position in enclosure	177	39	59
Feeding	54	11	18
Movements	130	28	48
Attention behaviors	115	25	38

reference to the labels within the conversations is not a particularly prominent area of comment. Conversations about likes and dislikes (emotive) and anthropomorphic and human–animal interactions (all attitudes) were higher among the groups with an adult rather than in the groups of children alone. Table 7.6 shows that in almost two-thirds of all conversations, visitors comment about aspects of the exhibit other than the animal. Visitors are looking carefully at the "stage" upon which animals are presented as well as at the animal specimens.

Understanding Terminology

Even though we have touched on the importance of understanding how visitors talk about animals (Chap. 6), here we will discuss the important of understanding the terminology that is used by zoo staff and visitors. There are the basic terms that adults use with children. However, if children are expected to learn the names and use them correctly, they need to hear the name used in the correct context (Brown, 1958). Adults, especially teachers preparing for a zoo visit, should be clear about their use of terminology. As stated in Chap. 6, young children learn animal terminology from their parents. Therefore, adults can aid children in learning new words and the word's appropriate usage. Adults can listen to how children apply the words they already know and help students relate those words to new concepts. However, a problem occurs when a simple explanation for something is used because the simplicity might cause the child more confusion rather than developing understanding.

Table 7.6 The main categories of comments about live animals among primary school groups at a zoo ($n = 459$) (Tunnicliffe, 1996)

Number and per cent of the conversation units containing at least one mention of the category										
	n	%	%*	n	%	%*	n	%	%*	n	%
Animal focused										
Body parts				280	61	61				
Head	57	12	20							
Sense organs	30	7	13							
Body	20	4	7							
Reproductive	3	1	0							
Excretory organs	11	2	4							
Locomotory	39	9	14							
Tail	21	6	8							
Size/shape	138	30	49							
Body coverings	95	21	34							
Life history ("baby")	53	12	19							
Other	19	4	7							
Behavior				301	66	66				
Excretory	6	1	2							
Movement	130	28	43							
Food related	54	12	18							
Sexual	8	2	3							
Parental	13	3	4							
Interacting/animals	23	5	8							
Position in enclosure	177	39	59							
Sleep/awake	28	6	9							
Other ("noise")	52	11	17							
Names				401	88	88				
Popular name	281	61	70							
Common name	72	16	18							
Recognize only	98	21	24							
Phylum/class name	33	7	8							
Order/family/genus	193	47	48							
Compare with self	23	5	8							
Compare inanimate	22	5	6							
Compare extinct	13	3	3							
Compare another animal	59	13	15							
Misclassification	33	7	8							

% indicates percent of all comments; %* indicates the percent of all those comments included in the next highest category. Thus, the 57 conversation units that include reference to the "head" represent 12% of all 459 conversation units and 20% of the 280 conversation units that refer to at least one body part

Because much of learning focuses on the communication of words, vocabulary is of paramount importance. Adults are skilled at using words and use words to name objects and to describe actions and feelings. However, do learners notice that the members of their society have several different names for the same thing?

This layered alternate taxonomy occurs when zoologist know that pet cats are scientifically called *Felis domesticus*, but when zoologist talk to young children about cats, instead of saying "cat," the scientist refers to them as "pussycats" (Motherese). At the zoo and at home, cats are identified by a distinctive word (familiar name) such as Sharma, Pharaoh, Tiggy, and Suki. Even more confusing for young children is when cat owners discuss their particular breed of cat by using terms such as Balinese, Birman, Chartreux, Cornish Rexes, Egyptian Mau, Peterbald, and Tonkinese. However, the words house cat (common name), Sharma (familiar name), and Peterbald (breed) will all refer to the same *Felis domesticus*. In addition to using layered alternate terminology in naming animals, we also use the same terminology to name body parts. For example, scientist may call the lower front of the human body part of the ventral surface or the abdomen. However, in conversations among nonscientists, the abdomen may be referred to as the stomach or tummy. When adults talk to children, they commonly use the term tummy. In other words, we instinctively select the term we use to name a body part based on the people with whom we are talking (Anglin, 1977). For example, mothers tend to use a basic term, such as tummy, when they talk with children (Motherese).

The term animal is commonly misused. A simplistic definition of "animal" is all living things which do not make their own food but have to catch or collect food (omnivores or herbivores). However, children and many adults use the term "animal" to refer only to mammals (Bell, 1981). This may happen because usually the first animals that children experience other than their own kind are mammals such as dogs and cats. Moreover, elementary teachers in the USA use the term "bugs" in reference to insects. In the 1970s, the UK Schools Council 5–13 Science Project coined the phrase "mini-beasts" to refer to small invertebrate animals because mini-beast is a simpler name than zoological terms. Children can use and understand technical terms when there is no alternative. For example, dinosaurs are called by their scientific names, but children remember them. Kindergarten children and those in the first year of formal schooling naturally name dinosaurs using the scientific names of the order or family such as Tyrannosaurus Rex or T-Rex (Tunnicliffe, 1996). We have not yet invented simple terms for these animals. Therefore, society has adopted part of the specific name to use as the everyday or ethnobiological name (Berlin, 1978). Is there an age at which teachers should ensure that they introduce a word used by scientists?

In the following conversation, kindergarten students (ages 4–5 years) use everyday terms and are moving toward the Americanized special name "brown bear."

Grizzly Bear
Mother: Look at the bears.
Boy: They are brown bears!

This conversation occurs between a docent and students. The docent is attempting to use everyday specific names while talking about the white tigers.

White Tiger
Docent: What does he look like?
Boy 1:   Lion!

Docent: Besides a lion?
Girl: A cat!
Boy: 2:   Bobcat!
Docent: A bobcat. Great!
Boy 3:   A house cat!
Boy 1:   A cougar!
Docent: A cougar! Okay!

Older students delight in finding the correct name using the exhibit labels and correcting their peers. This is a conversation between 3rd and 4th graders (aged 8 and 9 years).

Jaguar
Boy 1:   Jaguar!
Boy 1:   Hey, pussy cat.
Boy 1:   It's a jaguar and a panther not a pussy cat!

Many times, children will confuse the stages in an animal's life. The word "baby" may be used by children with the words "little" and "big" to describe an organism. A "little cat" is assumed to be the baby (Looft, 1971). Big cats are assumed to be the adult. Children look at the size of the animal first then pay attention to other observable features and interpret what they see in their own experiential terms. The 9-year-old boys in the following conversation are looking at lions with a zoo docent:

African Lions
Docent: Which one is the male and which one is the female?
Boy: That's the poppa back there. That big guy. That's the king. That one.
Docent: How do you know?
Boy: Because he is the biggest.

The following is another example in which children are observing small cats. Children visit this exhibit after they have observed larger cats. The children make assumptions about the animals based on the cat's size.

Ocelot
Girl: Ah, kittens, aren't they cute and small?
Boy: I've seen alley cats that big!

The following conversation takes place in an aquarium between two 10-year-olds and an adult. Because the remora is smaller than a shark, but the children assume the remora is a baby shark.

Remora
Girl 1:   What is it? A baby shark?
Girl 2:   It's a mommy shark and a baby shark.
Girl 1:   I'm cold, Patty. It's a baby shark.
Girl 2:   Yeah. They are babies. They are alive.

The animals named above go through incomplete metamorphosis, but caterpillars, darkling beetles, and frogs undergo complete metamorphosis. Normally, students are able to identify the organisms that undergo complete metamorphosis and realize that these are the "babies." However, animals that undergo incomplete metamorphosis are referred to as baby based on size.

Educators should be cognizant of the quality, form, and function of the conversations that occur during a zoo visit. In what topics are visitors interested when they view zoo exhibits? If Falk and Dierking (1992) are correct in their statement presented at the beginning of this chapter, analyzing the content of groups' conversations will provide information about visitors' interest and knowledge. The information that the zoo deems as important may not necessarily be those that are seen as important in the science curriculum. Moreover, the visitors' interest is likely to be triggered by what the visitors expect to see. Therefore, listening to, identifying, and understanding the features and attributes of the discourse that occurs during a zoo visit are important. The following chapter addresses the differences and similarities of family and school conversations.

References

Anglin, J. M. (1977). *Word, object, and conceptual development*. New York: W.W. Norton.

Bell, B. F. (1981). When is an animal, not an animal? *Journal of Biological Education, 15*, 213–218.

Berlin, B. (1978). Ethnobiological classification. In E. Rosch (Ed.), *Cognition and categorisation* (pp. 9–24). Hillsdale, NJ: Lawrence Erlbaum Associates Inc.

Biggs, J., & Collins, K. (1982). *Evaluating the quality of learning: The SOLO taxonomy*. New York: Academic.

Bliss, J., Monk, M., & Ogborn, J. (1983). *Qualitative data analysis for educational research: A guide to users of systemic networks*. London: Croom Helm.

Bloom, B., Hastings, J., & Madaus, G. (1971). *Handbook on formative and summative evaluation of student learning*. New York: McGraw-Hill.

Borun, M., Chambers, M., & Cleghorn, A. (1996). Families are learning in science museums. *Curator, 39*(2), 123–138.

Britton, J. (1970). *Language and learning*. New York: Penguin.

Brown, R. (1958). How shall a thing be called? *Psychological Review, 65*(1), 14–21.

Bruner, J. (1990). *Acts of meaning*. Cambridge, MA: Harvard University Press.

di Castri, F., & Younès, T. (1996). *Biodiversity, science and development. Towards a new partnership*. Wallingford, UK: CAB International.

Ellis, A., & Beattie, G. (1986). *The psychology of language and communication*. Hove, UK: Lawrence Erlbaum Associates.

Falk, J. H., & Dierking, L. (1992). *The museum experience*. Washington, DC: Whalesback Books.

Gee, J. P. (2010). *How to do discourse analysis: A toolkit*. London: Routledge.

Gilbert, J. K., Osborne, R. J., & Frensham, P. (1982). Children's science and its implications for teaching. *Science Education, 66*, 625–633.

Glaser, B. (2011). *Getting out of the data: Grounded theory conceptualization*. Mill Valley, CA: Sociology Press.

Halliday, M. A. (1980). An interpretation of the functional relationship between language and social studies. In H. K. Pugh, V. J. Lee, & J. Swann (Eds.), *Language and language use* (pp. 158–167). London: Heinemann Educational in association with Open University Press.

Hensel, K. (1987). *Families in museums: Interactions and conversations at displays.* Unpublished Ph.D. thesis, Columbia University Teachers College, New York.

Hymes, D. (1972). Towards ethnographies of communication: The analysis of communicative events. In P. Giglioli (Ed.), *Language and social context* (pp. 21–44). Harmondsworth, UK: Penguin Education.

Lemke, J. (1990). *Talking science: Language, learning and values.* Norwood, NJ: Ablex Publishing Corporation.

Looft, W. (1971). Children's judgement of age. *Child Development, 42*(4), 1282–1284.

Lucas, A. M., McManus, P. M., & Thames, G. (1986). Investigating learning from informal sources: Listening to conversations and observing play in science museums. *European Journal of Science Education, 8*(4), 341–352.

McManus, P. M. (1987). *Communications with and between visitors to a science museum.* Unpublished Ph.D. thesis, Chelsea College, University of London, London.

McManus, P. M. (1989). What people say and how they think in a science museum. *Heritage interpretation.* London, Belhaven Press.

Parker, L. C. (2007). *The use of zoo exhibits by family groups to learn science.* Unpublished dissertation, Purdue University graduate school, West Lafayette, IN.

Patrick, P., Matthews, C., & Tunnicliffe, S. (2011). Using a field trip inventory to determine if listening to elementary school students' conversations, while on a zoo field trip, enhances preservice teachers' abilities to plan zoo field trips. *International Journal of Science Education.* Available at: http://www.tandfonline.com/doi/abs/10.1080/09500693.2011.620035, 1–25, iFirst article.

Schegloff, E. A., & Sacks, H. (1973). Opening up closing. *Semiotica, 8*, 289–327.

Sinclair, J. M., & Coulthard, R. M. (1975). *Towards an Analysis of Discourse.* London: Oxford University Press.

Spicer, J. (1994). The exhibition: Lecture or conversation. *Curator, 37*(3), 185–197.

Tough, J. (1977). *The development of meaning.* New York: Wiley.

Tunnicliffe, S. D. (1993). *We're all going to the zoo tomorrow, zoo tomorrow: Children's conversations at animal exhibits at London and St. Louis Zoos.* Paper given at Visitor Studies Association Conference, Albuquerque, New Mexico.

Tunnicliffe, S. D. (1994). Why do teachers visit zoos with their pupils? *International Zoo News, 41*(5), 4–13.

Tunnicliffe, S. D. (1995). The content of conversations about the body parts and behaviours of animals during elementary school visits to a zoo and implications for teachers organising field trips. *Journal of Elementary Science Education, 7*(1), 29–46.

Tunnicliffe, S. D. (1996). Conversations within primary school parties visiting animal specimens in a museum and zoo. *Journal of Biological Education, 30*(2), 130–141.

Tunnicliffe, S. D. (2000). Conversations of family and primary school groups at robotic dinosaur exhibits in a museum: What do they talk about? *International Journal of Science Education, 22*(7), 739–754.

Vygotsky, L. S. (1978). *Mind in society.* Cambridge, MA: Harvard University Press

Chapter 8
School and Family Groups' Conversations

> Cultural/bio-conservatism is based on the recognition that humans are dependent upon
> the viability of natural systems. Critical thought and technological innovation that help
> carry forward ecologically sustainable cultural patterns are as important to cultural/bio-
> conservatism as are the wisdom based traditions handed down and renewed over genera-
> tions. (Bowers, 1997, p. 5)

School is not the only source for science learning (Bell, Lewenstein, Shouse, &
Feder, 2009) and is not named as a primary location for science learning (Falk &
Dierking, 2010). Moreover, even though organisms are a key part of the environment
and have an important part in children's lives, we do not have an extensive
understanding about what attracts children to animals. Identifying where people
learn about animals and the environment is important because their understandings
of the "environment are profoundly shaped by their attitudes towards, experiences,
and understandings of living organisms (Tunnicliffe & Reiss, 2000, p. 131)." Since
Louv's *Last Child in the Woods* was released in 2006, there has been an overall
acceptance of his analysis that children are out of touch with nature. However,
children are in touch with plants and animals that they see in their everyday lives
(Patrick & Tunnicliffe, 2011; Tunnicliffe, 2010) and have affective feelings that are
acquired from the culture, home, and school. Therefore, school visits and family
visits to zoos may be a missed educational opportunity in terms of learning new facts
and consolidating knowledge (Tunnicliffe, Lucas, & Osborne, 1997). This leads us
to ask the question, do school visits and family visits have similar educational roles?

Visitors visit the zoo with existing ideas, images, and information regarding
animals. During the discourse that occurs within the exhibit (Chap. 7), visitors
reveal their understandings and misconceptions about animals. Visitors do not
visit as *tabula rasa*, knowing nothing. Therefore, visitors' prior concepts are
important and must be elicited, considered, and analyzed. Visitors will use their
prior knowledge to construct their own meaning. In the following conversation,
the adult has prior knowledge that the golden lion tamarin is facing conservation
issues.

P.G. Patrick and S.D. Tunnicliffe, *Zoo Talk*, DOI 10.1007/978-94-007-4863-7_8,
© Springer Science+Business Media Dordrecht 2013

Tamarins

Adult: Look!
Child: It's like a little lion.
Adult: Yes. Golden lion tamarins. There are not a lot of those.

In the case of children, their knowledge, beliefs, and ideas may not make sense to parents and teachers (Driver, Squires, Rushworth, & Wood-Robinson, 1994), but children's ideas are ingrained and are a reflection of how children see the world. These ideas must be considered and should (1) form the starting point of the discourse concerning animals prior to the zoo visit, (2) present a lens through which zoos focus exhibit design (Chap. 5), and (3) influence how zoos utilize the experiential space (Chap. 5). In order to achieve these objectives, zoos must identify and understand the conversations juxtaposed in the exhibits' experiential space. Zoos should listen to the Visitor Voice and visitors should listen to the Zoo Voice.

The content and form of visitors' conversations are dependent on three main factors: (1) The situation in which the conversation occurs (Britton, 1970) influences discourse. (2) Discourse is influenced by the rationale for the visit. Families visit during their leisure time, but school groups visit for an educational purpose. (3) The composition of the groups will influence the conversation and the outcomes. Groups with children are divided into two main categories: school groups and family groups. Adults may be present in both categories of groups and may change the behavior of the children in the group. Because field trips are taken for educational reasons, school groups are expected to have a more focused conversational content about animal taxonomy and characteristics than family groups (Marshdoyle, Bowman, & Mullins, 1981; Tunnicliffe, 1994).

Zoos are places where various types of conversations occur. However, the most likely conversations are social ones which focus on the visitors and learning conversations (Lucas, McManus, & Thomas, 1986) which focus on the exhibit (Dierking, 1987; Hilke, 1989; McManus, 1987). The content of conversations that take place within a zoo exhibit varies little between family groups and that of school groups. The main variation is that school visits are predominately an affective experience. Children initiate many of the conversations and adults direct the exchanges surprisingly infrequently. The majority of comments are of an observational nature or a commentary of what is seen, and there is little effort by adults to help the children construct their knowledge or relate the concepts inherent within the exhibits. When the adult voice is heard, the adult asks questions within the discourse. In some conversations, children name and observe and the adult extends the observations through questioning. However, most conversations end at the identifying level (Borun, Chambers, & Cleghorn, 1996) and do not develop the available science teaching and learning opportunities. The lack of interaction between adults and children may be because the Zoo Voice provided through its exhibits is neither easy to read nor well told (Tunnicliffe et al., 1997). The school visit is a missed educational opportunity in terms of introducing children to new concepts through which the students can scaffold meaning. Both zoos and teachers should focus their efforts during the zoo visit on imparting new knowledge.

The content of conversations may reveal the rationale for the zoo visit, what the visitors talk about, and more. Conversations are used for a variety of functions, and particular functions have characteristic forms. To summarize the previous chapters, conversations have varied uses that are defined by the originator and the responses of those being addressed. School and family conversations consist of instruction, hijacking the attention of a member by another group member, and talk of a subject irrespective of the topic being addressed. Occasionally, philosophical debates between children will occur without the intervention of an adult. Adults facilitate observational opportunities for children in the group but leave the children to construct their own dialogue. Parents may relate the child's observations to prior knowledge or a context the parent understands.

The dialogue at animal exhibits *may* contain both elements of leisure and education, which forms a distinct *animal exhibit* style of dialogue. Recognizing the *zoological language* or *everyday language* that comprises the conversations of primary school-aged children with families or school groups is of significance. Zoo educators and classroom educators need a formal understanding of the extent to which teaching and learning dialogues are used in animal exhibits. The educators need to know if the animal engenders the same comments without the presence of an adult. Moreover, educators must be aware of how words are understood, acquired, and used in conversation and whether the experience of viewing the exhibits aids the visitor in acquiring biological knowledge. Research should pay attention to the differences in school groups and home-schooled children.

Family Groups

Chapter 7 describes conversations as a sociocultural activity that occurs as visitors move through the zoo and share discourse. The conversations that occur among families have been a major focus of museum research since the 1990s. In 2004, Ellenbogen, Luke, and Lynn recognized this rise in discourse analysis among families and stated that

> family learning research comes from the increasing popularity of sociocultural theory in learning research, and its suitability for museum research. A sociocultural perspective frames learning in and from museums as socially and culturally constructed through people's actions within a specific community of practice. A community of practice, such as a family, shares a set of values, vocabulary, understandings, and assumptions (Wenger, 1998). A person's actions and interactions are interpreted by members of their group (Green & Meyer, 1991; Gumperz, 1986), allowing them to construct meanings specific to the group through their conversations (Geertz, 1983; Green & Dixon, 1993). According to Edwards and Westgate (1994), "As we listen and as we talk, we learn what is necessary to know, do, and say in that area of social life or that setting and can display the competence necessary to be accepted as a member" (p. 15). It is largely through discourse that we come to identify ourselves as part of a community. (p. S50)

The family adults stimulate conversation by *directing* the child with instructions (Diamond, 1986) or *questioning* the child about what they see (Dierking, 1987).

Directing and questioning are the most frequently cited categories of verbal interaction within an exhibit. Family and school groups generate remarkably similar content in their conversations when they look at taxidermied specimens in a natural history museum and at zoos (Tunnicliffe, 1996). Ash (2003) has found that during informal education visits . . .

> parents used a variety of inquiry skills such as observing, questioning, hypothesizing, explaining, and interpreting as they talked to their children. In Gleason and Schauble (2000), parents were good at guiding their children's understanding in some skills—for example, in obtaining and planning evidence—but weak in helping children to interpret the evidence to draw conclusions. Their results suggest that perhaps 'adults may assume that children's understanding is the same as theirs' so that if 'an experimental outcome is observed by both, there is not need for discussion' (Gleason & Schauble, 2000, p. 45). Parents may assume that their children understand events as they do, or that science is just a matter of 'looking and seeing' (Driver, Leach, Millar, & Scott, 1996). (Ash, 2003, p. 156)

Studies of the form and function of children's conversations at home and at school show that the form of the conversation changes (Hughes & Cousins, 1988). At home, children ask questions and initiate the dialogue most often. However, at school, the teacher does the talking and usually initiates the dialogue (Tizard & Hughes, 1984). The conversations, within family groups at living animal exhibits (e.g., Hensel, 1987; Rosenfeld, 1980), consist of teaching dialogue and are generated by participants. Because these conversations are present in leisure visits, it would seem that the dialogue would be based solely on the social experiences. However, this is not the case. Group members encourage other group members, especially children, to acquire some information from the exhibits. The *ethos of school* affects the way in which parents talk with their children on the zoo visit. When parents visit the zoo with children, the parents may take on the identity of a traditional teacher and ask learning questions because the self-imposed teacher identity affects the parents' behavior (Falk, 2009). Home-schooled children are educated outside of the school experience and may shape their interests and enthusiasms during a zoo visit.

When family groups and non-field trip groups visit the zoo, they have a similar division of attention and discourse. In non-field trip groups, parents dominate the conversation at interactive exhibits in museums (Diamond, 1986), but during zoo field trips, children are allowed to view exhibits in peer groups without an adult. Therefore, understanding, identifying, and comparing the content of the various types of conversations that occur in zoo exhibits are important. Parallel conversations occur in which one person holds a conversation with several other people at the same exhibit. The parallel conversations that result during a non-field trip group visit may be characterized in the following ways:

- **Type A** *Adult–Children* The adult transmits or provides instruction or information to children in the group.

  Dad: Everybody look here! Soft-shelled tortoises. Can you see them?

- **Type B** *Adult–Child–Adult.* The adult begins a triadic dialogue in which they are asking questions or teaching.

Mother: It's a Rhinoceros Iguana. What group of animal is that?
Boy: A lizard thing.
Dad: It's a lizard, that's right!

- **Type C** *Child–Child.* The children may not be accompanied by an adult but do have their own unofficial personal conversational agenda. The child–child conversation is at a different level and may cover different topics, and many times, the interpretations and emphasis regarding the animals are different from adults.

Older boy (hippopotamus exhibit):   Look! You can see the break, that's refraction breaking the light.
Younger boy: Yes, cool!

- **Type D** *Child–World.* The child in the group broadcasts exhibit-related statements to no one in particular.

Boy: Hey! Look at this!

- **Type E** *Child Egocentric.* The child broadcasts their needs to no one in particular.

Boy: I'm getting all sweaty! I need a drink! Can we go?

- **Type F** *Adult–Adult.* Private agenda conversations in which two or more adults are looking after a group of children. The content of the discourse is often irrelevant to the exhibit's context but in some cases supports the verbal interactions.

Girl: It's a Toucan.
Mother: No it's not.
Mother's friend: It's got an itchy beak. I wish I could reach parts like that!
Boy: Mom! Mom!

- **Type G** *Adult–Child.* When an adult in the group is a mother or father to one of the children. Because of the family relationship between the child and adult, family conversations will intertwine with the *teaching talk* that occurs with the group's other children.

Boy 1 (son):   It's a giant iguana. It looks like my iguana.
Adult (father):   Don't bang on the glass!
Girl: It's looking at me!
Adult (father):   It is an iguana.
Boy 2:   It's turning its head!
Boy 1 (son):   It has black marks like Iggy.
Girl: Look at its tongue!
Adult (father):   I see its tongue.

- **Type H** *Adult Broadcast.* The adult broadcasts a statement out loud to the world in general, but the statement does not have any meaning.

Adult: Well I'll be darned!

The above conversational types can overlap within one group. For example, an *adult–adult* type and *child egocentric* type may occur simultaneously.

The conversational content of the discourse among families exposes the visitors' talk but can reveal more. The following conversation unveils the parallel that visitors draw between what they identify and their personal context:

Gorilla

Mother: You stand there and look.
Girl: Look! He looks sad. Look!
Mother: He is sad. As if to say, "What are you look looking at?" Isn't he?
Girl: Oh he looks like Granddad.
Mother: He does, doesn't he?

The mother and child watch the gorilla and seek to explain what they see within their own experience. They make sense of the expression on the animal's face by relating it to Granddad. The exchange also reveals a typical management style of parents who expect their children to look at the animals. However, during the visit, some parents engage in *teaching talk* with young children and "label"' new objects or manage their child's looking/observing in a similar way as teachers do in the classroom. For example, in the conversation below, the mother (10-year-old boy, 5-year-old girl) begins the conversation by sharing information from the signage and proceeds to ask questions about the characteristics that define birds.

Harpy Eagle

Mother: This is a harpy eagle.
Boy: Wow!! It's big.
Mother: It is one of the largest eagles in the world.
Girl: I want to see.
Boy: What is that on its head?
Mother: This is a bird?
Boy: What is on its head?
Mother: What do birds have that no other animals have?
Girl: What?
Boy: Feathers.
Mother: Very good!

In some family conversations, the form is managed by the mother, but she makes no attempt to teach. The discourse is an exchange of comments at the same level. The discourse may be started by the children with the mother joining in at a nonteaching level. Therefore, the discourse is only in a social exchange.

Okapi

Child: Look! It looks like a half zebra.
Mother: Yeah. It does.
Child: I like zebras.
Mother: I like zebras too.

Penguins

Girl:	They have spiral path (a reference to the exhibit).
Mother:	Penguins like the rain!
Dad:	I hate the rain.
Girl:	Why are they queuing?
Mother:	Perhaps they are early for lunch.
Girl:	Can we come back at 2:30? [There is a sign next to the penguins that states the penguins are fed at 2:30.]
Dad:	Yes!

The discourse that takes place at the penguin exhibit is between three people (mother, father, and child) and is an example of an elaborated triadic dialogue. Toward the end of the conversation, the girl initiates questioning and is answered twice, once by mother and once by father. The girl extends the conversation but does not close the conversation. Instead, the girl changes the content of the conversation by asking why the penguins are queuing. She poses a question and receives an answer. The answer given by her mother serves as a cue for her next question. The girl interacts with the exhibit signage and asks to return for the feeding. In turn, she influences the organization of the visit's pattern by asking to return to the penguin pool during feeding time. Her last question does not close the exchange but defers to the father which provides him an opportunity to become an active manager of the group.

In terms of quality of conversations in zoo exhibits, most of the dialogue is at a factual level formed from everyday observations and information acquisition. Few parents, teachers, or chaperones lead their children into relational thinking exchanges. However, conversations are diverse among family and school groups and become inverse triadic dialogue in which the child initiates the conversation (opportunity for teaching dialogue). The following is an example of an inverse triadic dialogue in which the child begins and ends the dialogue during a field trip:

Girl:	Oh ain't they lovely. I like them orange birds.
Teacher:	They are Scarlet Ibis.
Girl:	They look like flamingos, only smaller.

Adults attempt to point out ostensive biological features or behaviors to their children by calling their attention to examples with which the adult is familiar. For example, in the following discourse, the parent can see the animal breathing. However, the child is not listening because his attention is captured by a neighboring exhibit. When another animal hijacks the child's attention, the child is rendered "deaf" to the teaching efforts of his mother. This is a consequence of an exhibit with an active animal placed near an inactive animal.

Alligator and Tortoise

Mother:	See him breathing Nicholas? [Mother is pointing to the alligator.]
Child:	Look at that giant turtle! Look at that giant turtle! [Child is pointing to the tortoise in the next exhibit.]

Sometimes the present adult is not a part of the discourse that occurs and does not influence the construction of dialogue or learning. Children may debate the accuracies and inaccuracies of their knowledge without the intervention of a present adult. In this instance, the adult only intervenes with a management statement, suggesting that the view of the lions might be better from a different vantage point. In the African lion exhibit, the 7-year-old girls are accompanied by an adult, but the adult only uses management strategies and allows the children to construct their own dialogue and meaning.

African Lion
Girl 1:   I wonder how they feed them and what they eat.
Girl 2:   It's really high with a roof. Do they throw it over?
Girl 1:   No. They must throw it through the roof with sticks.
Girl 2:   Throw what?
Girl 1:   Throw it up.
Mother: I think there's more further round.
Girl 1:   We want to go further round.
Girl 2:   Yes. Let's go!
Mother: Yes. I think we'll see better round there. Come on!

During some rare occasions, *talking to learn* dialogue will occur. In *talking to learn* dialogue, children ask questions and parents provide answers based on the child's prior knowledge and relate what is seen to what children have seen before. In the Siberian tiger exhibit, the child asks about the net that covers the exhibit and the parent provides an answer, but did not relate it to the family cat "Timothy." Instead, the child notices the Siberian tiger rolling over and says he looks like Timothy (the family cat). This could have been an opportunity for the mother to interact in a dialogue that expanded the child's knowledge of the family cat to the larger cat family.

Siberian Tiger
Boy: Why have they a net over the top?
Mother: Because they are very strong jumpers and they could jump out. Aren't they?
Boy: Oh!
Mother: It's like the net over the garden at home. [The tiger rolled over onto his back.]
Boy: He looks like Timothy at home.

However, in the following exchange between family members in the African lion exhibit, the mother provides new information and seeks to aid the son in writing down the new information. This family has a spotter's guide (interpretive guide) to help them interpret the exhibits and they *talk to learn*. The interpretive guide aids the user in spotting and identifying zoo animals. The following discourse is an example of child identifying and the parents interpreting and applying information (Chap. 7):

African Lion
Boy: What's that? I don't think he's a lot like a horse. [Child is looking at information in spotter's guide.]

Mother:	Have you ticked this one off?
Boy:	Look Mummy, look Mummy!
Dad:	You can write down that you've seen the female too, because the female has no mane and is on the other side of him.
Boy:	You write it.
Dad:	If you make a noise, a dying noise, he might look at you.

School Groups

School groups are taken to the zoo to learn specific topics that the teacher has defined as relevant to the curriculum (Marshdoyle et al., 1981; Tunnicliffe, 1994). School subgroups consist of student only, teacher–student, and chaperone–student, who are carrying out educational tasks during a zoo visit and who have been effectively included in the educational content of the visit. The school subgroups should have a similar conversational content. However, there are some differences between the groups. Teacher-led groups make significantly more comments about the anatomy of the animals (size and color mostly) and ask more questions or give more pronouncements. Chaperone-led groups make more management comments (e.g., Come here!) and social comments (calling the name of someone in the group). Children-only groups have a freedom to express feelings that may be inhibited by adults. DeWitt and Hohenstein (2010) have found that

> the promotion of student autonomy, which is also supported by the decreased proportion of teacher evaluative statements, could certainly contribute to affective learning outcomes for these experiences, such as increases in self-confidence. Other researchers have suggested that museums and other out-of-classroom settings may have a valuable role to play in contributing to affective learning outcomes (Bell et al., 2009; Csikszentmihalyi & Hermanson, 1995; Rennie, 1994; Rix & McSorley, 1999; Wellington, 1990). Researchers have also called for students to be allowed more authority over their learning on school trips (Griffin, 2004). As we have shown here, teacher–student discourse on school trips tends to be more supportive of student autonomy than it is in the classroom. Thus, a worthwhile future research endeavor could investigate whether encouraging teachers to promote students' volunteering statements would be associated with increased student authority in the museum. (p. 468)

Adult-led school groups generate more knowledge source comments and questions and gave opinions more often than student-only groups. However, in the adult-led groups, both students and adults make knowledge source comments and ask questions. The field trip subgroups do not significantly differ in the frequency of "knowledge source" comments when compared with families, but the quality of the comments may be different. Student-only groups view the animals in an everyday observational manner and notice aspects of the exhibits. These are similar to those referred to by families, but children-only groups are less concerned with "management" comments and finding the animal. Children are more concerned with affective issues and those of authenticity in terms of it being alive. The learning that occurs in an informal setting relies on conversational cues like the cues used in formal learning situations, such as a word, image, or concrete item. The nature of the cue and its position in a learning sequence (introduction, middle, or consolidation) affects the interpretation of the item (Boulter, Selles, Tunnicliffe, & Reiss, 1997).

If the field trip is designed to help pupils identify animals and clearly focus on at least one of the minds-on levels of interaction and progress, then the form, function, and content of the subgroup's conversations should have similar characteristics. The form, function, and content of conversations depend on the situated learning that is taking place. If the students are learning, then the students will progress from identifying animals to interpreting their own knowledge of and experience with animals to scientifically interpreting animals. This takes more than one visit. Over numerous visits, students may apply scientific categorization and embedded knowledge to describe the animals in biological terms such as adaptations, anatomy, and behavior and finally interpreting and applying the knowledge gained. Conversely, if school visits are meant to have an affective, emotive, and social experience and are for making unstructured observations about the animals, then the content of the subgroups conversations will vary to some degree.

During a zoo field trip, children frequently view the accompanying adults as managers (Birney, 1988), and the children's interactions are much different when they are not accompanied by adults. Therefore, the presence of a chaperone has an impact on the way in which school groups behave during a field trip (Parsons & Muhs, 1994) and on how the students interact within the exhibit. When adults are present in the exhibit, the conversational patterns and content are different, and the adults' attitudes influence the students' opinions (ten Brink, 1984) and conversational content.

Teachers elicit relevant knowledge from students through conversation and respond to pupils' comments and attempt to share experiences with students in a meaningful way (Mercer, 1996). Teachers use dialogue to tell students about personal science experiences or to tell stories as a scientific explanation for a phenomenon. Teachers use dialogue to elicit differing opinions from students, clarify information, address misconceptions, suggest ideas, define expectations, and explain counterintuitive results. Educators use verbal interactions to aid learners in constructing meaning and understanding (Brown & Campione, 1996; Brown, Campione, Metz, & Ash, 1998). Moreover, educators use verbal interactions to educe relevant knowledge from students, respond to students, describe relevant learning events, and share experiences. The story sets the scene, creates a need for explanation, elicits differences of opinions from students, promises clarification, suggests ideas, creates expectations, and provides results. Educators provide further meaning by employing a variety of teacher–learner verbal interactions into their *storytelling*. The teacher–learner interactions include demonstration, practical investigation, discussion, and explanation. When teachers utilize explanation, they use relevant language and their knowledge of their students' learning levels to deliver a new vision of understanding. The teachers' explanations depend on their subject matter knowledge, personal experiences and resources, the nature of the subject matter, and the development of student understanding (Ogborn, Kress, Martins, &

McGillicuddy, 1996). Even though these ways of talking have been only studied in the classroom, they are equally applicable to encounters at animal exhibits that are part of a curriculum experience.

Children who have visited the zoo during a field trip remember parts of the visit that relate to the curriculum or actively involved them in some interaction. Moreover, children who have visited the same zoo are more likely to have memories of the visit (Wolins, Jensen, & Ulzheimer, 1992). Therefore, adults will have memories of past zoo field trips. These field trips will shape how the parents interact with children when they chaperone a group. When parents are chaperones during a field trip, parents take on the persona of the teacher that they remember from their childhood field trips (Tunnicliffe, 1995). This infers that the topics chaperones discuss with the children will be placed within a personal context of affective attitudes, episodic memories, and personal emotion, attitudes, ethics, and values. The personal context that chaperones bring also will influence the chaperones' management of the group, social comments, and exhibit access comments.

The information provided to children by an adult, particularly a teacher, can exert an influence over conversations even when the adult is not present. School has a powerful effect on students, even when they are out of school. In the following conversation at a mongoose exhibit, the dialogue between two 10-year-old girls during a field trip shows how the absence of an adult influences the conversation and how semantics influences certain languages. The children have some confusion over the meaning of an *animal* and *mammal*.

Mongoose
Girl 1:   Mongeese have nice little pink noses
Girl 2:   Which one is it?
Girl 1:   Those noses!
Boy 1:   We are mammals.
Girl 1:   You know Mr. Walsh Lynn?
Girl 2:   Yeah.
Girl 2:   Well, he said that we are mammals!
Girl 2:   Well, we are!
Girl 1:   Good job we are not animals!
Girl 2:   Look, they sleep on top of the rock!
Girl 1:   Where are they?
Girl 2:   Oh, look! They're coming toward us!

Moreover, when children who visit during a field trip are allowed to visit an exhibit without an adult, the children may make up their own stories to fit their observations. The following two discourses between four children shows that children will interpret what they see in everyday terms and will not use the biological knowledge they may have acquired at school:

Australian Frilled Lizard
Boy 1:   Carry on mate! Carry on mate! [Lizard is walking to pool.]
Boy 2:   Ah! Yes, he's going in. Yes! Yes!
Boy 2:   And does he like it!

Boy 1:   Yes, he's licking it; he's having a drink.
Girl 1:   God! Look how fat that one is.
Boy 2:   Yes, he's in. Hurrah!
Boy 2:   No, don't go that way! Now, you're heading in the right direction.
Girl 1:   Look at the gills on its head. [The child is referring to the folds on the lizards head.]
Girl 2:   He's so still isn't he? Like a statue.
Boy 2:   Look at the end of its tail.
Boy 1:   What do you want it to do?
Boy 2:   Go in the water.
Boy 1:   It would be fun if it slipped in.
Boy 2:   It nearly did at the beginning.
Boy 1:   Ah! He's coming . . .

Giraffe (several babies in the exhibit)
Girl 1:   They are tall, aren't they?
Girl 2:   Look at that giraffe?
Boy 1:   That's as tall as a house!
Boy 2:   They eat trees. They eat too much.
Boy 1:   That tree used to have a lot of leaves on it.
Boy 2:   I'd only come past his knees.
Girl 1:   How come they do not let the babies out?
Boy 1:   Because they would easily go over.

When children are taking part in a field trip, the children may be preoccupied with taking photographs to share with their family. This parallel agenda of taking photographs interferes with or prevents students from observing the animals. The following exchange between 12-year-olds observing Koalas provides a look at the interference an additional agenda can have during a visit:

Koala
Boy 1:   It's not real. It is a dummy.
Boy 2:   No, it ain't a dummy. It just moved its ears!
Boy1:   No flash! You are not allowed to use the flash!
Girl 1:   I think they are cute.
Girl 2:   He's not doing very much.
Boy 2:   Good job. I brought my dad's camera then.

Sometimes, taking photographs is a field trip assignment and meant to be used as a way to document the visit. However, instead of taking photographs, an incident will happen that attracts their attention. For example, in the following dialogue, the 11-year-olds are deflected from their original purpose of taking photographs to observing animal behavior. The children discuss and rationalize the animal behavior by relating the animal behavior to human behavior.

Otter
Girl 1:   I have got them both on a rock. [Talking about her photograph.]
Boy 1:   Aren't they lovely?

Boy 2:   I got them both at the top when they were running. [Talking about his photograph.]
Girl 1:   What's that pink thing? [Rubber toy used for enrichment.]
Boy 1:   Look, he's got a fish in his mouth.
Girl 2:   Urgh!
Boy 1:   Look! It's a bird.
Girl 2:   It's just a way of living.
Girl 3:   Well we don't go and get an otter and kill it and eat it, do we?
Girl 2:   No, but we kill cows and sheep and things.
Girl 3:   He's showing off now.
Boy 1:   He's pulling the guts out.
Girl 2:   Oh, thanks. I don't think I want to eat my sandwiches now.
Boy 1:   He's pulling the guts out of the birds. Jane, he's pulling the guts out of the bird!
Girl 2:   Well, we don't pull the guts out of a pig or something, do we?
Boy 2:   Oh, look! Here it is! Look!
Girl 1:   Which one is the one which was eating?
Girl 3:   I like that one best.
Girl 1:   I think the darker one is the male I like that one best.
Boy 2:   They all eat birds.
Girl 2:   He's eaten the head!
Boy 1:   He'll spare the eyes.
Boy 2:   He's trying to be a good boy.

The discourse among school groups follows a pattern similar to that of family groups with the majority of exchanges initiated by the child. Sometimes, during the discourse, the teacher will relate the animal to a school (family, if with a family member) activity to provide context and encourage a learner to talk more.

Wolves
Girl: Are they sleeping?
Teacher: Yes. Wolves are like dogs.
Girl: They won't kill us? Eat us?
Teacher: No, they don't kill anyone and eat them!
Girl: Yes, they do!
Teacher: What book have you been reading? Little Red Riding Hood I imagine.
Girl: Yes! The wolf.

The teacher may relate the animal and the child's observations to prior knowledge and understanding. New information may not be voiced, but sometimes, information provided at the exhibits is incorporated into the dialogue.

Python
Teacher: This looks interesting. Is it a boa constrictor or a python?
Girl: It looks like a python.
Teacher: Looks like you are right again. How do you know?
Girl: Because it is the color of the python. It's got the marks on its back. It's on the sign!

The following is an example of a classic triadic dialogue. In the following dialogue, the teacher is asking the child to think about the tiger:

Tiger
Teacher: Why do you think one is underneath and not on top (of the shelter)?
Boy: Because it's too hot.
Teacher: Good boy! He's in the shade at the moment because it's hot.

In a few instances, teachers begin the dialogue in an observational and declarative style, not the traditional teaching style of beginning the dialogue with a question. Some teachers use the inverse triadic dialogue to build on children's spontaneous observations. The teacher encourages the children to remember information and will sometimes initiate a teaching dialogue when there are no comments from the children. The teacher in the following discourse uses the story *The Billy Goats Gruff* to draw the child's attention to the animal. However, the teacher does not take the opportunity to ask thinking questions or stimulate biological thought. The following discourse begins and ends with the teacher in a classic triadic exchange:

Goats
Teacher: They're lovely. They're not the same color as the Billy Goats Gruff. What color were they?
Girl: Brown and black and white.
Teacher: That's right!

Additionally, discourse among a school group is more likely to be interrupted by management statements. These statements break the spontaneous pattern of dialogue as exemplified below and cause the dialogue to move away from cognitive interactions.

Baboon
Girl: I can see one's bottom.
Boy: Ha! Ha! Me too! Look!
Teacher: You stand here, next to me!

Occasionally, the adults in a school or family group do spontaneously follow up a child's statement with a directed educational question. The children name and observe the organism, and the adult extends the conversation by providing directed questioning. Within the inverse triadic dialogue below, the chaperone is seeking to initiate an information exchange between herself and the boy, even though the initial comment was uttered by the girl.

Dairy Cow
Girl: Cows! That's a cow
Boy: Cow. Perhaps they'll do a poo!
Chaperone: Do you know what we get from cows, Sydney?
Boy: Yes, milk!

Children's dialogue indicates that they do think conceptually. The following discourses are representations of children using their prior knowledge and mental models to make sense of what they observe. The following conversation among 8-year-olds represents an interaction in which children recognize that the elephant is covering its urine:

African Bush Elephant
Girl: Look, the elephant has done a wee. What a lot of wee. It must have had a lot to drink!
Boy: What is the other one doing?
Girl: It's treading in the wee!
Girl: It's covering the wee with sand.

Moreover, children's discourse also shows that they do recognize simple animal characteristics and are able to use this understanding correctly to identify animals. The following exchange between 5-year-olds is an example of children utilizing their mental models of giraffes to identify correctly an organism as a giraffe. However, the adult does not develop the opportunity for teaching science. The adult could give a summary of the conversation. Even though the children picked out salient features, the adult does not interpret the conversation into an informative narrative.

Giraffe
Girl: They look like giraffes!
Boy: They are brown.
Chaperone: What do they have on their heads?
Girl: Little horns.

Furthermore, children search for representations which allow them to understand what they see and make a connection. In the following exchange, the children recognize the sheep and make a physical connection but yearn to make a clearer connection to their prior knowledge. This connection becomes the recall of a children's nursery rhyme, which the boy sings to the sheep.

Border Leicester Sheep
Girl: Ah! It's soft.
Teacher: Don't touch the eyes or the mouth.
Girl: Is it Baa Baa Black Sheep?
Teacher: Are you going to sing to him then?
Girl: No!
Boy: I will. Baa Baa black sheep have you any wool?

Children make connections between their prior knowledge and experiences and the animal they are observing. The boy below recognizes the reptilian features of the iguana and relates them to his prior knowledge of dinosaurs' characteristics. The girl relates the iguana's skin to her knowledge of tights. However, the accompanying adult fails to ask follow-up questions and challenge the children. Even though the adult draws the children's attention to the iguana's diet, she did not ask the children if the iguana is an herbivore or a carnivore.

Boy 1:   Look! Dinosaurs!
Boy 2:   He's looking at us.
Girl: He's got tights on.
Adult: Has he? There is his food, too. Fruit and vegetables.

Within school groups, the problem usually becomes a process in which the children and the teacher make statements, but neither is listening. Each speaker is talking, but no one is listening to the other speak and each has a differing agenda at the time of speaking. In the following discourse, the child is seeking information, but the information the teacher provides does not answer the question. The teacher then proceeds to make management comments and ignores the boy completely.

Dromedary and Bactrian Camel
Boy: What one is the Egyptian (looking at label)?
Teacher: Their fur is all scruffy; they make lovely carpets and tents out of them.
           Jack! Do you mind not stripping off! That is the third time I have had to
           speak to you; come here!
Boy: Miss! Which one is the Egyptian?

Teachers also ignore or miss statements and questions that could lead to an educational conversation. In the following discourse at an iguana exhibit, several cues are provided by the children as they observe the characteristics of the iguana, but the teacher does not discuss the external anatomy of reptiles:

Iguana
Girl 1:   It's lost some of its spines along its back.
Girl 2:   Its skin looks like material.
Teacher: Yes.
Girl 2:   Its veins are sticking out.

However, some teachers do seize the opportunity to provide new information in relation to an animal's common name. In the following exchange, the teacher provides a name for the birds that are being viewed, but instead, the girl extracts from her own mental model. The child is not familiar with the scarlet ibis, but she does have a mental model of a flamingo. The teacher should expound on the child's understanding of the flamingo and compare the external anatomy of the scarlet ibis and the flamingo.

Girl: Oh, ain't they lovely. I like them orange birds.
Adult: They are Scarlet Ibis.
Girl: They look like flamingos, only smaller.

A teaching conversation would draw on the analogy of birds, ask the girl why she believes they resemble flamingos, and explore her grouping of birds.

If field trips are meant to provide meaningful educational experiences and the students are studying science as the focus of their visit, we should hear science process discourse. The science process discourse should consist of questioning, hypothesizing, predicting, observing, finding evidence, and evaluating. During the field trip, students should compare the characteristics and behaviors of different animals, seek patterns, identify adaptations, identify animals as omnivores or herbivores, and identify their habitats. Students should be naming animals and justifying their taxonomy. However, these conversations are seldom heard. Instead, the conversations that take place during a field trip are very similar to those of family groups. This suggests that zoo educators and classroom educators have a lot of work to do in developing the science educational value of zoo visits, if zoo field trips have educational objectives.

Despite the educational agenda which is given as a rationale for undertaking field trips (Marshdoyle et al., 1981; Tunnicliffe, 1994; Wolins et al., 1992) and the families' social agenda (Rosenfeld, 1980), the frequency of comments between the school groups and families is somewhat similar. There are a few differences between school groups and family groups. (1) Primary school children produce more emotive comments during a field trip (Tunnicliffe, 1995). School groups are more likely to use terms like Ugh! Ah! Oh! I love that, I hate that, and It smells but less likely to use terms like Look! Where is it? Look over there, and I don't see it. (2) School groups ask more questions. (3) School groups mention the characteristics of the organisms more often than family groups. However, in family groups and school groups, adults do not do the summonsing and do not control the Visitor Talk, but the children in the group do.

The predominant concepts of discourse among school groups are locating the animal, identifying the animal with a familiar name, and stating the dimensions of the animal. The discourse in which visitors name animals indicates that the visitor uses the names that they bring with them, but only 21% use the labels to aid them in naming the animal. Visitors who read the label use the common name acquired from the label but do not state or read aloud the scientific name or other information provided in the signage. Moreover, visitors' conversations do not incorporate information about conservation, diet, natural habitat, and geographical location, which are featured almost universally on zoo labels.

School groups may be grouped heterogeneously or homogeneously. The combinations do make some difference in the conversations that take place within the zoo exhibit, because male groups comment about certain aspects of the exhibit that girls do not. An analysis of 158 all-male conversations and 199 all-female conversations shows that male groups name animals more often and female groups are more likely to express emotive attitudes and comment about behavior (Tunnicliffe, 1998). In the following examples, the female groups name the animals but add emotive comments:

Ants
Girl 1:   Oh, look! Teddy bears.
Girl 2:   Giant ants.
Girl 1:   Look, it (food) is smothered in ants.
Girl 2:   It makes me itch.

Leeches
Girl: Is there anything in here?
Adult: Let's look. Oh, yes, there is a leech!
Girl: A leech ! Oh! Yes.

Cockroaches
Girl 1: Ugh! Uck!
Girl 2: Cockroaches.
Girl 1: I don't like any of them.
Girl 2: Hum.
Girl 3: They have eaten all the inside of the apple.

The discourse that takes place at the live cockroach exhibit shows the emotional response that occurs within a familiar context. The discourse provides an excellent example of all-female groups as they interact and generate significantly more emotive attitudes. The following example shows the discourse within an all-male group in which the males are concerned with giving the animal the correct name, but do not add emotive comments.

Okapi
Boy 1: Wow! Look! What is it?
Boy 2: It looks like a horse.
Boy 3: No, it looks like a zebra.
Boy 1: I think it is a horse.
Boy 3: The sign says okapi (mispronounced).

Males are more fact driven during discourse, while females are more concerned with their feelings, concerns, and relationship with the organism. In the following conversation, the females name the organism but spend much of their discourse making emotive comments:

Girl 1: Oh, aren't they cute?
Girl 2: Aren't they gorgeous?
Girl 3: Oh, my God!
Girl 2: Oh, I love meerkats.
Girl 1: Oh, look at that one. Aren't they cuddly? They're lovely.
Girl 3: That is cruel. I don't like that.
Girl 2: I like the big one.

Talking Science

As shown in the conversations in this chapter, teachers and chaperones help children and parents help their children to match what they see at the zoo with a memory or mental model. The conversations of school chaperone–student subgroup are similar to those of family groups, with few small differences. Conversations among children without an adult produce a higher number of naming comments. The presence of an adult focuses the children's observations on the animal and evokes more emotive and affective comments. Chaperoned groups generate more management and social comments than other school subgroups (Parsons & Muhs, 1994). This is compatible with the chaperone's managerial role, but the chaperone's role is not equated with that of the teacher. Students perceive teachers as adults with status, because they keep students on task and expect and maintain standards of behavior. However, few teachers and chaperones effectively manage the exchange of cognitive learning dialogues. Teachers and chaperones need to consider how conversations during a zoo visit attribute to curriculum content and how the visit is a valuable teaching tool.

A superficial look at the animals is acceptable during a family visit, but if school children are visiting as part of the curriculum or to introduce or reinforce science learning, then a superficial look is not acceptable. Scientific concepts are developed during instruction and discourse between the learned and the learner (Roschelle, 1996). When teachers organize field trips, they should help the students enter the zoo with meaning. A definite educational objective must be stated by the teacher or the group leader that focuses the attention of the students. Most likely, the majority of chaperones do not possess biological knowledge other than general knowledge about the animal's characteristics. Therefore, teachers should prepare chaperones by briefing them about the visit and what scientific knowledge is expected to be discussed. If chaperones are not prepared, students will not receive the educational interactions to which they are entitled. However, teachers may brief chaperones, but the lack of focus on the organism's anatomical structures and characteristics may indicate that the chaperones did not share the educational objectives of the visit. When chaperones are prepared and do share the educational objectives of the visit, the children will be more likely to construct an understanding of biological and zoological science and taxonomy.

When children are brought to the zoo, their prior knowledge and the reason for the visit influence the type of discourse that transpires. If a young child is brought to the zoo on their first encounter with organisms, very little discourse may occur, because the visitor is at a familiarization level. The amount of informational exchange that occurs may be minimal during a first encounter. If the visit is part of a formal learning agenda such as a field trip, then a high degree of teaching by the teacher or chaperones should occur. During an educational visit, the conversations should reflect a cognitive discourse. Science-related discourse should reflect a use of previous knowledge, science content, and relate form to function in non-anthropomorphic terms.

Zoos are responsible for encouraging the adults and children who visit the zoo to *talk science* and share their observations. Lemke (1990) provides a definition of *talking science* that zoos should employ when developing educational programs. Lemke states that

> Learning science means learning to *talk* science. It also means learning to use this specialized conceptual language in reading and writing, in reasoning and problem solving, in guiding practical action in the laboratory and in daily life. It means learning to communicate in the language of science and act as a member of the community of people who do so. "Talking science" means observing, describing, comparing, classifying, ana-

lyzing, discussing, hypothesizing, theorizing, questioning, challenging, arguing, designing experiments, following procedures, judging, evaluating, deciding, concluding, generalizing, reporting, writing, lecturing, teaching in and through the language of science. (p. 1)

Understanding the conversations that take place within zoo exhibits is a baseline upon which zoos and their education departments can construct meaningful interpretation. The story presented by zoos is not easy to read and may not be well told. Zoo professionals must understand that most teachers and adults with school groups appear to possess little more additional knowledge about the specimens than the children themselves. Therefore, the teachers and adults are not able to develop the conversational content to a higher cognitive level than the children. The adults emphasize the features that the children spontaneously notice. The attributes children identify and the names used by children appear to be representative of their prior everyday knowledge of animals. Children are not spontaneously interested in the diet of the animal, its geographical origins, or its conservation issues. This understanding will provide a starting point so that children can construct personal and zoological meaning from their existing knowledge. If zoos identify the discourse that occurs within their collection, they could use the information to design interpretative material and educational packets.

References

Ash, D. (2003). Dialogic inquiry in life science conversations of family groups in a museum. *Journal of Research in Science Teaching, 40*(2), 138–162.

Bell, P., Lewenstein, B., Shouse, A. W., & Feder, M. A. (2009). *Learning science in informal environments people, places, and pursuits.* Committee on Learning Science in Informal Environments. National Research Council of the National Academies. Washington, DC: The National Academies Press.

Bell, P., Lewenstein, B., Shouse, A. W., & Feder, M. A. (2010). *Learning science in informal environments people, places, and pursuits.* Committee on Learning Science in Informal Environments. National Research Council of the National Academies. Washington, DC: The National Academies Press. www.nap.edu.

Birney, B. (1988). Criteria for successful museum and zoo visits: Children offer guidance. *Curator: The Museum Journal, 31*(4), 292–316.

Borun, M., Chambers, M., & Cleghorn, A. (1996). Families are learning in science museums. *Curator, 39*(2), 123–138.

Boulter, C. J., Selles, S. E., Tunnicliffe, S. D., & Reiss, M. J. (2011, April). *Pupils' responses to cues from the natural world: A cross-cultural study using multiple analytic perspectives.* Paper presented at the National Association for Research in Science Teaching, Orlando, FL.

Bowers, C. A. (1997). *The culture of denial: Why the environmental movement needs a strategy for reforming universities and public schools.* Albany, NY: State University of New York Press.

Britton, J. (1970). *Language and learning.* New York: Penguin Books.

Brown, A. L., & Campione, J. C. (1996). Psychological theory and the design of learning environments: On procedures, principles and systems. In L. Schauble & R. Glaser (Eds.), *Innovations in learning: New environments for education* (pp. 289–325). Mahwah, NJ: Erlbaum.

Brown, A. L., Campione, J., Metz, K., & Ash, D. (1998). The development of science learning abilities in children. In A. Burgen & K. Harnquist (Eds.), *Growing up with science: Developing early understanding of science* (pp. 156–178). Dordrecht, The Netherlands: Kluwer Academic.

Csikszentmihalyi, M., & Hermanson, K. (1995). Intrinsic motivation in museums: Why does one want to learn? In J. H. Falk & L. D. Dierking (Eds.), *Public institutions for personal learning* (pp. 67–77). Washington, DC: American Association of Museums.

DeWitt, J., & Hohenstein, J. (2010). School trips and classroom lessons: An investigation into teacher-student talk in two settings. *Journal of Research in Science Teaching, 47*(4), 454–473.

Diamond, J. (1986). The behavior of family groups in science museums. *Curator, 29*(2), 139–264.

Dierking, L. D. (1987). *Parent–child interactions in a free choice learning setting: An examination of attention directing behaviour.* Unpublished Ph.D. thesis, University of Florida, Gainesville, FL.

Driver, R., Squires, A., Rushworth, P., & Wood-Robinson, V. (1994). *Making sense of secondary science: Research into children's ideas.* London: Falmer Press.

Driver, R., Leach, J., Millar, R., & Scott, P. (1996). *Young people's images of science.* Philadelphia: Open University Press.

Edwards, D., & Westgate, D. P. G. (1994). *Investigating classroom talk.* New York: Falmer Press.

Ellenbogen, K., Luke, J., & Dierking, L. (2004). Family learning research in museums: An emerging disciplinary matrix? *Science Education, 88*(Supplement 1), S48–S58.

Falk, J. H. (2009). *Identity and the museum visitor experience.* Walnut Creek, CA: Left Coast Press.

Falk, J. H., & Dierking, L. D. (2010). Questioning the school-first paradigm: School is not where most Americans learn most of their science. *Informal Learning Review, 105*, 2–5.

Geertz, C. (1983). *Local knowledge: Further essays in interpretative anthropology.* New York: Basic Books.

Gleason, M., & Schauble, L. (2000). Parents' assistance of scientific reasoning. *Cognition and Instruction, 17*, 343–378.

Green, J., & Dixon, C. (1993). Introduction to "Talking knowledge into being: Discursive and social practices in classrooms. *Linguistics and Education, 5*(3&4), 231–239.

Green, J., & Meyer, L. (1991). The embeddedness of reading in classroom life. In C. Baker & A. Luke (Eds.), *Towards a critical sociology of reading pedagogy* (pp. 141–160). Philadelphia: John Benjamin.

Griffin, J. (2004). Research on students and museums: Looking more closely at the students in school groups. *Science Education, 88*(Suppl.1), S59–S70.

Gumperz, J. (1986). Interactive sociolinguistics in the study of schooling. In J. Cook Gumperz (Ed.), *The social construction of literacy* (pp. 45–68). London: Cambridge University Press.

Hensel, K. (1987). *Families in Museums: Interactions and conversations at displays.* Unpublished Ph.D. thesis, Columbia University Teachers College, New York.

Hilke, D. D. (1989). The family as a learning system: An observational study of families in museums. In B. H. Butler & M. B. Sussman (Eds.), *Museum visits and activities for family life enrichment* (pp. 101–129). New York: Haworth Press.

Hughes, M., & Cousins, J. (1988). The roots of oracy: Early language at home and at school. In M. Maclure, T. Phillips, & A. Wilkinson (Eds.), *Oracy matters* (pp. 110–121). Buckingham, UK: Open University Press.

Lemke, J. (1990). *Talking science: Language, learning and values.* Norwood, NJ: Ablex Publishing Corporation.

Louv, R. (2006). *Last child in the woods. Saving our children from nature-deficit disorder.* Chapel Hill, NC: Algonquin Books.

Lucas, A. M., McManus, P., & Thomas, G. (1986). Investigating learning from informal sources: Listening to conversations and observing play in science museums. *European Journal of Science Education, 8*, 341–352.

Marshdoyle, E., Bowman, M. L., & Mullins, G. W. (1981). Evaluating programmatic use of a community resource: The zoo. *Journal of Environmental Education, 13*(4), 19–26.

McManus, P. M. (1987). *Communications with and between visitors to a science museum.* Unpublished Ph.D. thesis, Chelsea College, University of London, London.

Mercer, N. (1996). The quality of talk in children's collaborative activity in the classroom. *Learning & Instruction, 6*(4), 359–377.

Ogborn, J., Kress, G., Martins, I., & McGillicuddy, K. (1996). *Explaining Science in the Classroom*. Milton Keynes, Philadelphia: Open University Press.

Parsons, C., & Muhs, K. (1994). Field trips and parent chaperones: A study of self-guided school groups at the Monterey Bay Aquarium. *Visitor Studies: Theory, Research, and Practice, 7*, 57–61.

Patrick, P., & Tunnicliffe, S. D. (2011). What plants and animals do early childhood and primary students' name? Where do they see them? *Journal of Science Education and Technology*. Invited article and Special Issue: Early Childhood and Nursery School Education. Available at: http://www.springerlink.com/content/e27121057mqr8542/, 1–25, iFirst article.

Rennie, L. J. (1994). Measuring affective outcomes from a visit to a science education centre. *Research in Science Education, 24*, 261–269.

Rix, C., & McSorley, J. (1999). An investigation into the role that school-based interactive science centres may play in the education of primary-aged children. *International Journal of Science Education, 21*(6), 577–593.

Roschelle, J. (1996). Designing for cognitive communication: Epistemic fidelity or mediating collaborative inquiry. In D.L. Day & D.K. Kovaco (Eds.), *Computers, Communication and Mental Models* (pp. 13–25), London: Taylor & Francis.

Rosenfeld, S. (1980). *Informal learning in zoos: Naturalistic studies on family groups*. Unpublished Ph.D. thesis, University of California, Berkeley, CA.

ten Brink, B.L. (1984). *Fifth grade students attitudes toward ecological and human issues involving animals*. Unpublished Ph.D. thesis, Austin, University of Texas at Austin, TX.

Tizard, B., & Hughes, M. (1984). *Young children learning*. London: Fontana.

Tunnicliffe, S. D. (1994). Why do teachers visit zoos with their pupils? *International zoo news, 41*(5), 4–13.

Tunnicliffe, S. D. (1995). The content of conversations about the body parts and behaviours of animals during elementary school visits to a zoo and implications for teachers organising field trips. *Journal of Elementary Science Education, 7*(1), 29–46.

Tunnicliffe, S. D. (1996). The relationship between pupil's ages and the content of conversations generated at three types of animal exhibits. *Research in Science Education, 26*(4), 461–480.

Tunnicliffe, S.D. (1998). Boy talk: Girl talk–Is it the same at animal exhibits? *International Journal of Science Education, 20*(7), 795–811.

Tunnicliffe, S. D. (2010, March). First Engagement with the environment. Children are in touch with nature. *People and Science*, p. 25.

Tunnicliffe, S. D., Lucas, A. M., & Osborne, J. F. (1997). School visits to zoos and museums: A missed educational opportunity? *International Journal of Science Education, 19*(9), 1039–1056.

Tunnicliffe, S. D., & Reiss, M. J. (2000). Building a model of the environment: How do children see plants? *Journal of Biological Education, 34*(4), 172–177.

Wellington, J. (1990). Formal and informal learning in science: The role of the interactive science centres. *Physics Education, 25*, 247–252.

Wenger, E. (1998). *Communities of practice*. New York: Cambridge University Press.

Wolins, I. S., Jensen, N., & Ulzheimer, R. (1992). Children's memories of museum field trips: A qualitative study. *Journal of Museum Education, 17*(2), 17–27.

Chapter 9
The Zoo Voice: Zoo Education and Learning

Most zoological parks continue to fail to go beyond superficial entertainment toward instilling greater appreciation of animals among children, while most learning of animals in school appears to be so divorced from direct experience with animals and the natural environment that little basic knowledge results (Kellert, 1985, p. 31).

Since Kellert's study in 1985; Wagoner and Jensen (2010) have found that children do gain knowledge of animals during a zoo visit. However, their study also

highlights the crucial role of variables outside of the direct context and motivations surrounding the zoo visit itself. The cultivation of pre-visit representations of animals, habitats, and the environment occur over an extended period of time through the influence of multiple sources, including formal education and mass media. Education within the zoo must interact with such pre-existing ideas in the process of visitors' development of a new understanding of animals and their environments (pp. 73–74).

Why Visit Zoos?

Previous chapters have established the historical development of zoos and education, built a rationale for zoos, discussed exhibitry, and defined the Visitor Voice. Yet the question remains, why should people visit zoos in this technological age of multimedia and vicarious representations of the living world? Zoos present opportunities for interactions between people and live animals and promote biological science (Wagoner & Jensen, 2010). When people visit zoos, there is a natural engagement that occurs between the visitor and the exhibits. This interaction may be biologically based or not.

People visit zoos mainly to see animals. Parents, relatives, friends, schools, and youth groups take children to the zoo to see animals. Some parents make pilgrimages to show their children certain animals. For example, a mother and father at the Isle of Wight Zoo in England stated they brought their child to the zoo to see the lions "before it was too late." The parents were concerned that in this time of

P.G. Patrick and S.D. Tunnicliffe, *Zoo Talk*, DOI 10.1007/978-94-007-4863-7_9, mass extinction, their child might not have another chance to see a lion. However, people have various other reasons for visiting the zoo. The zoo is also a cultural landscape (Hallman & Benbow, 2006) and is formed by the culture in which the zoo is located. The zoo forms a context in which family relationships are pursued (Hallman, Mary, & Benbow, 2007). Some visitors do not attend the zoo because they have a real interest in animals. Instead, some visitors view the zoo as a place to meet family members. For example, some separated or divorced parents use the zoo as a safe place to meet the children from the relationship. Families visit zoos while they are on vacation. Some visitors want to see the zoo's architectural design. The Lubetkin Penguin pool, the Giraffe House, and the Snowdon Aviary at the London Zoo and the Breakfast Pavilion at the Schonbrunn Zoo are examples of famous architectural designs that lure visitors. The London Zoo's famous architecture has inspired the London Zoo and the Civic Trust to produce a booklet about the zoo's architecture. Moreover, zoos are providing venues for personal celebrations and cultural celebrations. Christmas includes a nativity scene with animals such as a reindeer and a donkey. Halloween, which has become a commercial festival, is celebrated during Boo at the Zoo. Zoos are hosting corporate events, weddings, social gatherings, and birthday parties, but visitors attend for the event not the animals.

School groups ostensibly attend to address stated curriculum objectives. However, teachers admit that the zoo visits are important because the zoo is a safe place where children can practice some independence, and the visit gives children a special treat at the end of school year (Tunnicliffe, 1994). Even though the zoo visit must have a curriculum focus, the sentiment of controlled adventure is still there. Therefore, children taking field trips may regard the zoo visit as a break from the normal routine of the classroom and may be most excited about the journey and gift shop. However, the field trip should have a different agenda that revolves around learning outcomes and cognitive development.

Zoos provide formal education during on-site classes that are based on a predetermined curriculum and informal opportunities in which children are able to determine their own learning. For example, zoos hold summer camps for children and adults may attend art classes. Home schoolers are a growing audience, and in the USA, zoos (e.g., Lincoln Park Zoo) are beginning to cater to home schoolers by providing courses for the home school educators and opportunities to view the animals (Graison, 2010). Moreover, some zoos, such as the Cincinnati Zoo and the North Carolina Zoo, have secondary schools on site. The London Zoo has special general classes for younger children, zoological classes for older students, and zoological science lectures for adults. The adult lectures are followed by an optional dinner in which the diners can meet the speaker. Today, even the informal learning that takes place at the zoo is planned to some extent. Currently, there is an emergence of nature play in specific areas of the zoo where children are provided opportunities for constructive, messy play. These play experiences are important because some children do not have access to mud play and climbing rocks (Wagner, Becker, & Fulk, 2011). Animal care also is becoming a fundamental concept upon which conservation biological knowledge and positive attitudes toward animals and the environment are built. The Brookfield Zoo has established a pet kitchen and a pet responsibility area in the Hamill Family Play Zoo. Zoos are recognizing their importance in bridging the gap between the visitors' urbanized world and that of nature.

Prior Knowledge and Learning

As stated in Chap. 8, visitors do visit the zoo with prior knowledge. People have everyday experiences with nature when they fish and hunt, walk in the park, walk to school (Tunnicliffe, 2010), and watch a sunrise. In addition to everyday experiences in nature, people visit science and nature centers, zoos, aquariums, botanical gardens, nature reserves, and planetariums (Bell, Lewenstein, Shouse, & Feder, 2009). Therefore, zoo visitors bring prior knowledge, previous experiences, sociocultural influences, and formal teaching experiences that influence their personal pre-visit agenda (Falk, Moussouri, & Coulson, 1998). Even when a visitor is part of a group that has an overarching reason for the visit, the visitor's personal agenda affects their interest during the zoo visit, their interpretation and recall of the exhibits (Anderson, Piscitelli, & Everrett, 2008), and their interactions and conversations with other group members.

As determined in prior chapters, the discourse that occurs among visiting groups shows that children voice particular aspects of the animals. Therefore, zoos should listen to the discourse that occurs during a visit and use the information to develop interpretation and educational programs. The zoo's interpretation should begin with the terms visitors use and find familiar. Educational opportunities in zoos should begin at the point of interest and understanding of the visitor. The zoo's educational approach should assist visitors in constructing personal meaning. The zoo should cognitively develop the mental models and knowledge that visitors bring to the visit. Education in zoos should be a constructive process for visitors, which means zoos should not lecture to the visitor.

Prior knowledge is an important aspect to consider when zoos develop their educational programs. Children begin to learn about animals and their immediate environment during their infancy. They learn to recognize different types of animals and their basic names (Rosch, Mervis, Gray, Johnson, & Boyes-Braem, 1976). Certainly for most children, animals form a significant part of the world around them, whether as wildlife, pets, or zoomorphic toys. Therefore, familiar animal names form a large part of a child's vocabulary. Many early words come from identifying images of animals on household items such as furnishings, wallpaper, clothing, and soft toys (Gatt & Vella, 2003).

The earliest detailed work on children's conceptual development in biology has been carried out by Piaget (1929). Piaget states that there are five stages in the development of the concept of living: (1) Stage 0 (age 0–5 years). No concept of living. (2) Stage 1 (age 6–7 years). Things that are active in any way, including falling or making noise, are said to be living. (3) Stage 2 (age 8–9 years). All things that move, and only those, are said to be living. (4) Stage 3 (age 9–11 years). Things that appear to move by themselves, including rivers and the Sun, are said to be living. (5) Stage 4 (over 11 years). Only animals and plants are said to be living. Once children are able to identify plants and animals and understand they are living, they use a range of attributes to decide whether a specimen is an animal or not. Younger children rely on the criterion of movement (Osborne, Wadsworth, & Black, 1992), and older children rely on other criteria such as nutrition (Lucas, Link, & Sedgwick, 1979). However, when children decide on how different animals can be named, classified, and grouped, children of all ages consistently use anatomical features (Trowbridge & Mintzes, 1988; Tunnicliffe & Reiss, 1999a). Similarly, when children name, classify, and group plants, they use anatomical features most often. However, older children (11- and 14-year-olds) increasingly use habitat features (Tunnicliffe & Reiss, 1999b). Learning is crucial for the acquisition of biological concepts. Learning takes place principally when two conditions are satisfied. First, the learning takes on personal meaning to the learner, and second, the learning takes on a sociocultural context (Vygotsky, 1962). This view is challenged by Chomsky (Crystal, 1987) who argued that languages share certain universal properties despite their huge variability. These properties include the existence of rules from which all the grammatical sentences in a language can be derived. Children, according to Chomsky, are born with a "language acquisition device." Possession of this language acquisition device enables a child to make its own meaningful utterances based on what she/he has heard. For example, in English, a child hearing plurals such as houses, cars, and cats will derive (unconsciously) the rule that plurals are formed by the addition of an "s." This is why young children may say words like mouses and sheeps, even though they have never heard them.

Reiss (2005) states that:

> All human learning takes place as a result of interactions between an evolved brain and a sensed environment and the human brain is perhaps the most impressive of all the products of evolution. Stephen Mithen (1996) has argued that the contemporary human mind can be envisaged as having evolved over time rather as a small mediaeval cathedral might have grown through the addition of chapels. Around a core "general intelligence" structure, chapels of "technical intelligence," "linguistic intelligence," "social intelligence" and "natural history intelligence" were added. Natural history intelligence contains at least three subdomains of thought: that about animals, that about plants and that about the geography of the landscape.

If the human mind contains templates that have evolved to be more receptive to certain environmental stimuli, then children have an innate ability to learn about animals and plants, their features, and their environments. Gelman and Markman (1986) taught 3- and 4-year-old children new facts about objects. For example, they told children that small brown snakes lay eggs. The children were then presented with pictures of a cobra, a small brown worm, and a cow and asked whether the newly learned fact (small brown snakes lay eggs) describes the object in the picture. Gelman and Markman found that the children drew more inferences based on category membership than on perceptual appearances. For example, the children were more likely to state that the cobra laid eggs than that the small brown worm did, despite the latter's greater superficial resemblance to the original small brown snake.

The central importance of a social environment for learning has been emphasized by Vygotsky (1962), who recognized that effective learning is a social endeavor and influenced by adults and the particular culture in which the person lives. Effective teaching and learning is a bidirectional dialogue that occurs between the adult (teacher) and the child (student). Social environments are places for learning in terms of the discourse that takes place between the learner and her/his peers. Therefore, peer interaction enhances the development of logical reasoning through a process of active cognitive reorganization that results from cognitive conflict (Forman and Camden, 1985). The space around an animal exhibit is the place where logical reasoning can occur between adults and children and between children and children. This exhibit space is the zone of experiential space (Scheersoi & Tunnicliffe, 2009). The zone of experiential space is the distance between the actual developmental level as determined by independent problem-solving and the level of potential development as determined through problem-solving under adult guidance or in collaboration with more capable peers. The importance of peer communication is easily revealed when children's conversations are recorded (as discussed in Chaps. 6, 7, and 8). For example, when 6-year-old children are presented with bottled ecosystems of brine shrimp (*Artemia salina*), their spontaneous conversations (Tunnicliffe & Reiss, 1999b) express their understanding of what they see and the ideas they share with peers and the teacher. In seeking meaning, the pupils attempt to match what they see, which is unfamiliar, to something they already know. In so doing, the children employ metaphors drawn from their prior knowledge. The prior knowledge provides part of the interpretative framework that the pupil constructs individually and as a group. Biology teaching provides many opportunities for such collaborative learning, but all too often teachers are suspicious of extended pupil conversations, worried that too much talking may result in misbehavior.

There is a universal knowledge about animals that is called folk biological cognition (Atran & Medin, 2008) and is based on what the child sees in their everyday environment. Different cultures feature animals in various ways. In some cultures, local animals are part of the lives of the inhabitants, and the children from these cultures have an ecological understanding superior to that of urban children in more developed countries (Bang, Medin, & Atran, 2007). However, in some cultures, certain animals are associated with myths, such as bats and spiders being used for Halloween decorations. Both spiders and bats are phobic animals (Davey et al., 1998) and are commonly represented in horror films as dangerous. Moreover, many people hold biological misconceptions about bats and spiders. Bats are commonly considered birds and spiders are regarded as insects (Prokop, Kubiatko, & Fanoviová, 2007; Strommen, 1995; Trowbridge & Mintzes, 1985, 1988). These culturally influenced stories or alternative conceptions about bats and spiders have a significant relationship with children's negativistic attitudes toward these animals. Researchers in Slovakia (Prokop & Tunnicliffe, 2008) used two questionnaires with nearly identical items for identifying attitudes to bats and spiders in a sample of school participants ($N = 196$) aged 10–16 years. The data show that children (especially girls) show more negative attitudes toward spiders in comparison with bats.

Tomkins and Tunnicliffe (2007) have found that when children are given choices about the natural objects they would like to talk about, the children pick (1) animals with actions and behavior rather than unresponsive specimens; (2) items of novel nature or appearance; (3) items with which children have some prior familiarity; (4) aesthetic attributes of color, shape, texture, feel, weight, size, etc.; and (5) objects eliciting some affective feeling, emotional engagement, and/or anecdote. Based on these findings, zoo and classroom educators should provide children with opportunities to enhance their observational skills with hands-on opportunities like nature tables. They should select a wide range of familiar objects/organisms that will develop the children's knowledge about their own natural environment. Children will make careful observations if encouraged and allowed sufficient time. These observations are basic to the development of scientific understanding and scientific inquiry. Children will *read* objects for cues to understanding, and such cues engender questions and children's greater interest. Objects have more value when children are able to make an affective or emotional link to the object. Children may link their observations to myths and stories. Educators should explore these links with the students guiding the other students to compose reality from fantasy. Classifying, identifying, pattern seeking, observing, and exploring often are seen as fundamental to a wider set of process skills (Watson, Nixon, Wilson, & Capage, 1999). Observation is a universal activity that is not confined to science and may be dismissed as a very basic skill, but should not be disregarded as just looking.

Teachers and parents facilitate a child's learning about animals with differing types of conversation. As discussed in Chap. 7, the content of conversations that take place in exhibits also may be categorized into four distinct categories: (1) exhibit access, (2) exhibit focus, (3) management focus, and (4) social interactions. However, learning access conversations also occur. Learning access conversations happen in addition to the other four categories and embrace the learning functions that take place between adults and children. Learning access conversations are perceived to be the point of visiting the zoo. Learning discourse varies (Chin, 2007) and is used by zoo educators and visitors to aid students in constructing meaning. Adults may use inquiry techniques, an important part of science learning, in which they ask children to justify their observations and deductions and consider if the evidence for their assertion is valid. Language is important when teaching and learning using science inquiry. Students have to formulate explanations and evaluate their selection of evidence to support their claim. This process is referred to as argumentation. It is seldom heard in the Zoo Voice or Visitor Voice. However, occasionally, hints of scientific inquiry do appear in the discourse that occurs between a child and an adult. The following exchange is between a parent and an 8-year-old boy.

Cobra
Boy 1:   That's a Cobra.
Boy 2:   Yes.

Boy 1:   No it is not after all.
Father: Why have you changed your mind?
Boy 1:   Because it has not got those things on it [pointing to his neck and imitating the hood]. We saw them on the model over there.

When children ask for the name of an animal, the adult can encourage children to justify their observations with reasons or evidence. For example, if a child asks "Is that a lion?" instead of answering "yes" or "no," the adult should respond by asking what evidence and knowledge makes the child think that the animal is a lion. What is the child's evidence for the categorization? The following conversation begins a discourse in which an 8-year-old child identifies the animal incorrectly, but instead of using questioning techniques to determine why the child mislabeled the animal, the teacher states the name. However, she does provide corrected information.

Chimpanzee
Boy: Monkeys!
Teacher: No they are chimpanzees. They haven't got tails.
Boy: Ah! Where are they going? [Animal was climbing in the enclosure] Their hands and feet are just like ours! Look!

Out of school experiences are an important source of science literacy. Children develop a theory about natural phenomena before they experience any formal teaching (Driver, Squires, Rushworth, & Wood-Robinson, 1994). Therefore, real experiences are paramount such as playing in nature (Vadala, Bixler, & James, 2007). Significant formative experiences in the wild often lead to involvement in environmental conservation in adulthood (Palmer, 1993; Tanner, 1980). Children must come to know and love the natural world. This stage is a vital first step that must occur before children will be concerned with the care of the natural world. People in the USA and Europe experience similar childhood events that shape their childhood association with animals. These childhood associations affect adults' interest in environmental concerns, activities, and ethics (Chawla, 2006, 2007). Zoos are becoming more aware of this phenomenon and are striving to design educational activities and other opportunities that may contribute to these values in visitors (Leinhardt & Knutson, 2010). Zoos now have a mission to enhance public understanding of wildlife and wildlife conservation. Zoos are working to identify and implement plans in which their facility can encourage the visitors to watch and investigate animals in their local environment (Street, 2010).

Looking at animals requires recognition and categorization skills. Therefore, children's careful observation of natural objects is basic to their development of science understanding and scientific thinking. Klahr (2000) broadly describes four scientific thought processes that occur in a stair-step pattern: (1) inquiry, (2) analysis, (3) inference, and (4) argument. These four steps are spontaneous for children, reinforced through observation and experience, and stimulated by imagination and a developing sense of causality. Just as nature table objects allow children to make an affective or emotional link (Tomkins & Tunnicliffe, 2007), biofact tables permit children to express their prior experiences when looking at animal artifacts. At a biofact table, children will rely on understanding gained through firsthand experiences.

Educators should ethically and safely cater to the kinesthetic aspects of observing, which are particularly important to young children. We recognize that children, with their spontaneous interest, will readily link their own observations with myths and stories. Teachers should use the children's stories to explore students' misconceptions and link them with the organism being viewed to compose reality from fantasy. Active learning is much more than manipulating materials. To truly learn, students must be allowed and encouraged to discuss what has happened, why their results might vary, how the current investigation connects to previous investigations, and what other questions might be explored. However, in formal education, student learning is measured by success on a mandated exam. Under this guise of mandated testing, students *have to learn* the *word* that will get the highest mark, but do not have to possess a true understand of the concept. In the present assessment-driven climate, teachers feel unable to teach biology in a holistic way and are not able to concentrate on the joys of learning and discovery. Teachers do not feel that they are able to spend time building deep conceptual understandings. Zoo educators also are feeling the strain of teaching to the test. Patrick and Middleton (2011) have analyzed zoo websites and found that 49% of the AZA-accredited zoos indicate there is professional development available for teachers, 70% have available activities aligned with state standards, 32% offer teachers pre-visit and post-visit packets, and 47% of the websites mention the availability of classroom kits that are aligned with state or national science standards. All too often, teachers and zoo educators are occupied with teaching the information that students need to pass high-stakes tests. The emphasis on testing has forced teachers to focus on *how to* rather than *what does this mean?* (Tunnicliffe & Ueckert, 2007).

If students are to connect ideas and develop deep understandings over time, then learning goals should have a coherent emphasis with connections in and across subject areas and grade levels. Formal and informal educators need to understand the previously stated concept and offer free-choice learning. Free-choice learning research has demonstrated that educational outcomes are often indirect and constructed by the zoo visitor as much as they are influenced by the zoo's educational staff (Clayton, Fraser, & Saunders, 2009). Recent studies in science education have highlighted the importance of the social aspects of classroom learning. Previous studies of zoo visitors have primarily focused on the family and other leisure visitors, but now research is becoming more focused on teachers and field trips. Understanding the discourse among all groups that visit the zoo should be the foundation upon which zoo educators base their talks, interventions, exhibit designs, and labeling.

Zoo Education

Zoos aspire to educate visitors and researchers view zoos as preeminent institutions for informal learning (Rennie & McClafferty, 1996). However, the public views zoos as a place to see animals and spend time with their family, friends, and schoolmates. However, zoos view themselves as educational institutions that support scholarship and programs that extend conservation knowledge to the public (Patrick, Matthews, Ayers, & Tunnicliffe, 2007; Patrick, Matthews, Tunnicliffe, & Ayers, 2007). Classroom educators expect field trips to be scientifically focused on animal taxonomy and characteristics, conservation, and environmental concerns.

Zoo education is an umbrella word that encompasses the declarative approach of providing information in a didactic style, invitational labeling, interpretation, physically interactive exhibitry, aiding visitors in constructing an understanding of animals, developing the visitors' prior knowledge, and addressing the visitors' misconceptions. Zoo education should lead the visitors into interactive learning. Our understanding of the psychological processes involved in learning in zoos is still in its infancy, but learning about animals in a zoo is different from learning in a classroom.

In formal education environments:

> The Education system makes strenuous efforts to engage children at as young an age as possible to win a lifetime's commitment to their agenda. Crucially zoos do not have to try hard to do this. There is a primal demand on the part of children to learn about, and be splendidly immersed within the animal world. In time this early exposure can lead into a lifelong commitment to the environment, 'green' principles, and engagement in science training and careers. (Kemsley, 2010, p. 20)

Informal, free-choice learning research has shown that educational outcomes are often indirect and constructed by the visitor. Learning may not be determined by the activities designed by zoo educators, information provided by the zoo, signage, or live presentations, because learning includes emotional dimensions and personal meaning-making that occur in the social context of the visit. Visitors may become enthusiastic about protecting individual animals and species as they learn about the animals (Clayton et al., 2009) and develop a relationship with the animal. Visitors often want to find out more about an animal and its conservation plight as they begin to feel a relationship and connection develop with the animal. The animal channels the social interactions and group narratives that take place during a visit.

If teaching in informal environments occurs with the progression of the learner in mind, the formal and informal educator must think about the learner and carefully plan the continuity of ideas and messages that are being delivered (Driver et al., 1994). Therefore, zoos must understand the learner and aid the visitor in constructing their own biological concepts. In Chap. 8, we discuss how *talking science* is important during the zoo visit. Even though *talking science* is an important aspect of science classrooms, talking science may not occur during an encounter at an exhibit. Educational opportunities in zoos should begin at the point of interest and understanding of the visitor and progress into *talking science*. Because of the lack of ability to *talk science* and establish a taxonomy vocabulary, zoos in South Africa are teaching zoology as a holistic subject (Gordon, 2010).

Zoos have marketing departments that maintain a variety of strategies that encourage visitation such as adopting animals, summer and overnight camps, junior keepers, nonpaying internships, special holiday activities, and behind-the-scenes tours. Moreover, some zoos are engaged in wildlife tourism, but the tours do not prompt a change in the environmental attitudes of the participants (Ballantyne, Packer, Hughes, & Dierking, 2007; Packer, Ballantyne, & Falk, 2010). Zoos traditionally focus on the cognitive aspects of learning, but the dimensions of self-efficacy and social identity provide a look into motivational domains. Therefore, zoo educators should understand the psychological processes involved in learning about animals by considering the visitor's self-efficacy and social identity. Understanding visitors' self-efficacy and social identity will allow zoo educators to tap into the visitor's personal and motivational domains. Self-efficacy is a person's belief in their ability to achieve a goal or an outcome. People who have a strong self-efficacy are more likely to take on challenges. Self-efficacy is important to zoos, because people who believe they are capable will be more likely to interact with exhibits and hands-on activities. Social identity is the way a person views their place in society and is based on associations and interactions with other people, behavior, and personal values and ethics. Social identity concerns an individual's sense of self and includes their visiting group and comments. Social identity places zoos within the culture of the visitor and their perceived importance of zoos.

Undergraduate and graduate students do have ideas about zoos and their educational content. Undergraduate and graduate students at Cambridge University were asked about their interest and attitudes toward a zoo visit before and after visiting a zoo (Tunnicliffe, 2001). Thirty-four percent of the respondents expected to observe educational aspects during visits, and 57% expected to observe some aspects of zoology. However, only 5% anticipated gaining information about biological conservation, and only 2% anticipated to learn about animals' welfare. The post-visit data shows a positive response to the zoo in general and that college students had noticed a conservation message. The outstanding event that the students described was a presentation by animal keepers of the ruffed lemur, green-winged macaw, Quaker parakeet, eagle owl, and ferret. The keeper talks conveyed a strong conservation message. Prior to the zoo visit, the college students' main expectations were educational and zoological. Post zoo visit, the students' focus widened to encompass the zoo's facilities, nature of the buildings, and the exhibits' suitability for children. The college students thought that the "Animals in Action" show was an excellent educational event and did a wonderful job of illustrating the zoo's mission, "conservation in action." Preservice elementary teachers from the USA, who listened to and recorded elementary students' conversations during a zoo field trip, determined that elementary school groups did not discuss animal information or conservation. After listening to the conversations that take place during a field trip, the preservice teachers realized that they did not fully understand the amount of preparation involved in conducting a school field trip. The study endorsed the view that preservice teachers should be required to visit organizations that are field trip destinations and learn to organize and prepare field trips (Patrick, Matthews, & Tunnicliffe, 2011).

Education is considered part of the zoos' mission, and zoos are anxious to show that education does occur. In Chaps. 7 and 8, we discuss the significance of analyzing the discourse that takes place in zoo exhibits and the importance of using the content of conversations as an indicator of visitors' interests (Falk & Dierking,

1992). Visitors' comments indicate how successful an exhibit is in eliciting visitors' relevant experiences, prior knowledge, and memories and how effectively visitors make links between their existing knowledge and the exhibit. However, whether or not the visitors truly learn new information and their observations are retained is another issue. Understanding learning in zoos is increasingly complex and depends on the knowledge and agenda with which the learners visit. Zoos can identify the success of an exhibit by determining if the exhibit elicits relevant memories from the visitors and assists exhibit observers in linking their prior knowledge (Falk, 2009). However, the memories and prior knowledge that are evoked by the visit are not the only important aspect of a zoo visit. If zoos are having an impact, then the visitors' observations at the exhibit will be retained (Wagoner & Jensen, 2010). An analysis of spontaneous conversational content provides a broad indicator of the psychological involvement of visitors and a clue to the rationale for the visit. Analyzing the discourse of visitors' conversations during the zoo visit does provide a broad indicator of the psychological involvement of visitors.

Visiting a zoo and being a member of a social group are not sufficient in bringing about learning during a zoo visit. Additional interpretation and guidance during the visit is required. Even though providing or facilitating an affective experience is not sufficient in identifying oneself as an educational organization, Kellert (1985) suggests that educational efforts among children 6–10 years of age should be focused on the affective realm and emphasize emotional concern and sympathy for animals. The affective domain could serve as the starting point in an educational program for young children and lead to learning about the other aspects of zoological science. Using an affective orientation to learning could be the starting point for the educational objectives which zoo curricula demand. Therefore, if zoos intend to produce effective educational objectives, the zoo and classroom educators who scaffold the learning that will take place during the zoo visit should consider (1) the stages of a visit, the characteristics of a visit, and the attention of pupils during the visit; (2) the anatomy, characteristics, and behaviors of the animals that pupils are likely to notice spontaneously; (3) the colloquial or everyday names students use to identify animals and any scientific names the zoo would like to have used; and (4) the concepts that the zoo wants visitors to acquire. If these conditions are not met, visitors will continue to talk about the same things and visit for leisure purposes.

Many zoos have stands or tables similar to nature tables. The stands are operated by volunteers or docents as they are termed in the USA. The volunteers have been trained in the information that can be elicited from the biofacts (pelts, snake sheds, bones, feathers, etc.). Visitors stop to ask questions about the biofacts, and sometimes the volunteers ask passing visitors an engaging question to draw them into a dialogue. Some of the interactive experiences allow the visitor to touch or interact with the biofacts. For example, at a polar bear enclosure, visitors are invited to place their hand inside a lard-filled, double-bagged ziplock bag. The bag with the visitor's hand inside is then placed in cold water and the other hand is place in cold water, so the visitor can feel the effect of the polar bear's "blubber." Engaging with visitors interactively and focusing on one animal species send the visitor a message about conservation and make the visitor more likely to support the work of the zoo financially. Zoo Atlanta visitors who had an interactive experience with the elephant demonstration and biofact program said they would be more likely to actively support elephant conservation (Swanagan, 2000).

Science teaching has been influenced by constructivist psychology and is based on Kelly's personal construct theory (Bannister & Fransella, 1971), which recognizes that education is as an *active* process between the learner and the teacher. Vygotsky (1962) recognized the importance of conversations in learning and pointed out that *spontaneous concepts* are developed by internalized and personal conversations. *Scientific* concepts are developed through formal school-type dialogues that take place between the child and the teacher. Effective teaching and learning dialogues can be stimulated by various tasks such as problem-solving, designing taxonomies, and group discussions that lead to explanations for physical phenomena. This understanding of the nature of learning has profound implications for both the visitors and zoos because learning cues need to be provided which help the visitor construct further scientific meaning for themselves about the animal specimens. Understanding the verbal and physical interactions that occur within nature or natural settings has profound implications for visitors and zoos, because zoos need to provide learning cues that aid the visitor in constructing scientific meaning through discourse. Zoos should play a greater role in the development of the visitors' knowledge about animals. Careful planning of the interactions visitors have with the animals should result in the visitors' heightened awareness of nature deficit and environmental threats. By listening to visitors' conversations, zoos will be able to determine if visitors are reaching Kahn and Kellert (2002) six stages of cognitive maturation. Kahn and Kellert's six stages of cognition are:

Knowledge: Understanding facts and terms and applying this knowledge to the articulation and presentation of ideas, developing broad classificatory categories and systems, and recognizing causal relationships

Comprehension: Interpreting and paraphrasing information and ideas and extrapolating these understandings to other contexts and circumstances

Application: Applying knowledge of general concepts, ideas, and principles to various situations and circumstances

Analysis: Examining and breaking down knowledge into elements and categories and discerning underlying structural and organizational relationships

Synthesis: Integrating and collating parts or elements into patterned, organized, and structural wholes and identifying and generating understandings of relationships and interdependencies

Evaluation: Rendering judgments about the functional significance and efficacy of varying elements and functions based on careful examination of evidence and impacts (Kahn & Kellert, pp. 62–63)

The six stages of cognitive maturation move sequentially from understanding facts to problem-solving. Visitors will not achieve the last level of cognition without being led through the lower levels of cognition.

Because 86% of the school children who visit zoos are of primary age (Braus & Champeau, 1994), educating teachers is an important aspect to be considered when zoo educators work to develop successful field trips. If zoos educate teachers about field trips and help them prepare students to visit the zoo, they will increase children's knowledge of animals and improve conservation attitudes. In the past 30 years, literature has emphasized the need for long-term professional development activities that revolutionize the educational setting (Loucks-Horsley et al., 1990). In a study that has been completed by McLaughlin, Smith, and Tunnicliffe (1998), a brief teacher zoo training course (i.e., 2 days near the beginning and end of the academic year with time during the intervening months for classroom application of the training) has had a positive effect on both the instructional practices of teachers and the educational outcomes for students. When teachers are introduced to the zoo's resources through workshops, they are more likely to use the zoo-related activities and information. Whereas untrained teachers, who do not attend the zoo workshop, are unequivocally less likely to include as much zoo-related instruction in their teaching. The trained and untrained teachers have very different zoo field trips.

The trained teachers' students are more likely to demonstrate a significantly higher level of writing skill when describing the zoo visit. Moreover, the students include science concepts like habitat when discussing their view of a zoo exhibit and their writing has a central theme, an appropriate word choice, and a wealth of the relevant vocabulary. The activity reflects students' writing skills and their mastery of the content about which they are writing. Therefore, students of the trained teachers, who have been instructed in the zoo's educational content and resources, reflect positive results because of the zoo-related instruction. Professional development in an informal science setting that takes into account Kahn and Kellert (2002) six stages of cognitive maturation can positively affect the teachers' attention to field trip design and the students' knowledge level and understanding and application of new knowledge. Zoo-related training allows teachers to learn about additional resources and extend their use of already established teaching techniques. Teacher workshops are provided by zoos around the world as part of teacher professional development. Knowing what is at the site, feeling comfortable with the zoo's layout, and learning to develop a field trip advanced organizer have a positive result on the outcome of the field trip (Falk, Martin, & Balling, 1978). Pre-secondary teachers are often more comfortable with the teaching of language arts (Weiss, 1997), but providing zoo education can increase the teachers' attention to science.

To provide teachers with solid inquiry-based activities, zoo educators must also be familiar with the stages of the zoo visit. The stages that occur during the zoo field trip influences what the visitor attends. There are five main stages of concentration and focus that occur during a zoo visit and are defined by the visitors' behavior: (1) Orientation, (2) Focused Looking, (3) Leisure Looking, (4) Completion, and (5) Consolidation and Extension. Visitors begin with an Orientation stage in which they look around, look at the map, and find their way. For school groups, the duration of this stage can be shortened if the teacher orients students and adults to the layout of the zoo prior to the zoo visit. These advanced organizers may consist of Powerpoints, Prezis, videos, timetables, information about the gift shop visit, when and where lunch will be, and similar housekeeping arrangements. Teachers can give students maps and pose various problem-solving questions using orientation skills.

For example, teachers can ask students "What route would you take to reach the Reptile House from the Bird House?" The use of maps prior to the visit will allow students and adults to think about the day of the visit and about what they will see and do. Use welcome sheets to ask them what they want to see and what they think they will enjoy. Provide pictures of animals and ask some questions about the animals' name and characteristics. In other words, key the visitor into what they will see and observe prior to the zoo visit. If teachers do not provide students with an Orientation phase, their attention may be focused on where they are going, lunch time, toilets, and other thoughts which take them from focusing on animal. The Orientation stage can be thought of in three parts: (1) in school, (2) on the journey, and (3) arrival at the zoo.

The Focused Looking stage follows the Orientation stage and occurs in 4 ascending levels of complexity. Level 0 is when the focus is on looking at the animals but in a leisurely manner. Level I occurs when the group focuses on a particular theme and looks at certain animals. Visitors do not just wander around aimlessly looking at animals from a particular region or looking for red or large animals. The discourse that occurs does include questioning and answering but in an information checking manner. Level II is when there is some recognizable difference in content and form between the leisure visitors' and school groups' conversations. The topics discussed relate to the curriculum and its skills such as reading and spelling. Level 3 is when education is identifiable. New knowledge acquisition occurs and discourse progresses in terms of content and form. Teachers ask learning questions of the pupils and vice versa, but the way in which the answers are handled and the questions phrased show educational dialogue, not everyday communications. Observations are related to previous knowledge. Level 4 is when groups or individuals raise questions, plan how to find the answer, make observations, and interpret their data using prior knowledge and firsthand observations. In the focused looking stage, people focus on set tasks or look in a concentrated manner at the exhibits. The Focused Looking phase may involve visitors in an educational activity provided by the zoo. Expecting students to be involved in a focused task throughout the visit is unrealistic.

After a period of time, visitors' concentration will wane and they will move into the Leisure Looking stage. The visitors' comments and observations will be similar to those of non-field trip visitors. When teachers and zoo educators do not provide educational content for the visit, school groups will maintain the Leisure Looking stage throughout the visit and may not move through the Focused Learning stage. The Completion stage occurs toward the end of the visit. The group is concerned with visiting the gift shop and gathering together and preparing for the journey home. The completion stage can be extended by the provision of zoological word searches, quizzes, crosswords, and thought-provoking questions about what they liked best and what they learned. These may be completed as a group on the bus home. Once back in school, there is the Consolidation and Extension stage in which teachers use the experiences from the actual visit to develop further concepts and consolidate the learning that took place during the visit (Tunnicliffe, 1999b).

There is a tripartite aspect of zoo educational visits that can be likened to a zoo sandwich, in which the learning experience at the zoo is the filling and the pre-visit and post-visit work form the bread. In other words, there are three stages in designing a field trip. (1) *Pre-visit activities* access prior knowledge process and content such as language arts, speaking and listening, reading and writing, mathematical knowledge, geographical knowledge, and science learning. If the visit is arranged for pupils to focus on biodiversity, the pupils may well practice making and using simple keys, learning Linnaean classification, and the major taxonomic groupings of animals before they visit. (2) *During-visit activities* ask students to apply previous knowledge in a new situation, identify new information, build on previous work, and work on conservation projects. (3) *Post-visit activities* are completed back at school and consist of a review, discussion, and sharing conclusions. Misconceptions can be challenged and remedial action taken through discussion.

A visit to the zoo does not occur in isolation and should be planned as an integral part of the students' learning experiences. The adults with any group also learn using the cognitive, social, and affective domains, which may have a longer impact on their conservation related knowledge (Briseno-Garzon, Anderson, & Anderson, 2007). The visit to an animal collection is an opportunity for the institution to enhance the knowledge of children, adults, and family members who did not come but may hear about the visit. The zoo visit is an opportunity for learners to build on their prior knowledge of animals by observing the animals' structure, taxonomy, behavior, needs, and conservation issues. The zoo visit is a missed opportunity when zoo educators and classroom educators fail to capitalize on the learning potential. Therefore, zoo educators and classroom educators must work in partnership to develop educational activities.

References

Anderson, A., Piscitelli, B., & Everrett, M. (2008). Competing agendas: Young children's museum field trips. *Curator, 51*(3), 253–273.

Atran, S., & Medin, D. L. (2008). *The native mind and the cultural construction of nature*. Boston, MA: MIT Press.

Ballantyne, R., Packer, J., Hughes, K., & Dierking, L. (2007). Conservation learning in wildlife tourism settings: Lessons from research in zoos and aquariums. *Environmental Education Research, 13*(3), 367–383.

Bang, M., Medin, D. L., & Atran, S. (2007). Cultural mosaics and mental models of nature. *Proceedings of the National Academy of Sciences of the United States of America, 104*(35), 13868–13874.

Bannister, D., & Fransella, F. (1971). *Inquiring man: The theory of personal constructs*. Harmondsworth, UK: Penguin.

Bell, P., Lewenstein, B., Shouse, A. W., & Feder, M. A. (2009). *Learning science in informal environments people, places, and pursuits*. Committee on Learning Science in Informal Environments. National Research Council of the National Academies. Washington, DC: The National Academies Press.

Braus, J., & Champeau, R. (1994). *Windows on the wild: Results of a national biodiversity education survey*. Washington, DC: World Wildlife Fund.

Briseno-Garzon, A., Anderson, D., & Anderson, A. (2007). Adult learning experiences from an aquarium visit: The role of social interactions in family groups. *Curator, 50*(3), 299–318.

Chawla, L. (2006). Learning to love the natural world enough to protect it. *Barn, 2*, 57–58.

Chawla, L. (2007). Childhood experiences associated with care for the natural world: A theoretical framework for empirical results. *Children, Youth and Environment, 17*(4), 144–170.

Chin, C. (2007). Teacher questioning in science classrooms: Approaches that stimulate productive thinking. *Journal of Research in Science Teaching, 44*(6), 815–843.

Clayton, S., Fraser, J., & Saunders, C. D. (2009). Zoo experiences: Conversations, connections, and concern for animals. *Zoo Biology, 28*, 377–397.

Crystal, D. (1987). *The Cambridge encyclopedia of language*. Cambridge: Cambridge University Press.

Davey, G. C., McDonald, A. S., Hirisave, U., Prabhu, G. G., Iwawaki, S., Jim, C. I., Merckelbach, H., de Jong, P. J., Leung, P. W., & Reimann, B. C. (1998). A cross national study of animal fears. *Behaviour Research and Therapy, 36*, 736–750.

Driver, R., Squires, A., Rushworth, P., & Wood-Robinson, V. (1994). *Making sense of secondary science: Research into children's ideas*. London, UK: Falmer Press.

Falk, J. H. (2009). *Identity and the museum visitor experience*. Walnut Creek, CA: Left Coast Press.

Falk, J. H., & Dierking, L. (1992). *The museum experience*. Washington, DC: Whalesback Books.

Falk, J. H., Martin, W. W., & Balling, J. D. (1978). The novel field-trip phenomenon: adjusting to novel settings interferes with task learning. *Journal of Research in Science Teaching, 15*, 127–134.

Falk, J., Moussouri, T., & Coulson, D. (1998). The effect of visitors 'agendas on museum learning. *Curator, 41*(2), 107–120.

Forman, E., & Cazden, C. (1985), Exploring Vygotskian perspectives in education: The cognitive value of peer interaction. In J. Wertcsh (Ed.), *Culture, communication, and cognition: Vygotskian perspectives* (pp. 323–347). New York, NY: Cambridge University Press.

Gatt, S., & Vella, Y. (2003). *Constructivist teaching in primary school*. Malta: Agenda.

Gelman, S. A., & Markman, E. M. (1986). Categories and induction in young children. *Cognition, 23*(3), 183–208.

Gordon, L. (2010). 2010 international year of biodiversity: Zoos role in teaching zoology as a holistic subject. *International Zoo Educators, 47*, 9–15.

Graison, S. (2010, September). *Zoo Atlanta's Home School Academy*. Paper presented at the Association of Zoos and Aquariums, Houston, TX.

Hallman, B., & Benbow, M. (2006). Canadian human landscape examples. *Canadian Geographer, 50*, 256–264.

Hallman, B. C., Mary, S., & Benbow, P. (2007). Family leisure, family photography and zoos: Exploring the emotional geographies of families. *Social and Cultural Geography, 8*(6), 871–888.

Kahn, P. H., Jr., & Kellert, S. R. (2002). Responsive management. In P. H. Kahn Jr. & S. R. Keller (Eds.), *Children and nature: Psychological, sociocultural, and evolutionary investigations* (pp. 50–95). Cambridge, MA: The MIT Press.

Kellert, S. R. (1985). Attitudes towards animals: Age-related development among children. *Journal of Environmental Education, 3*, 29–39.

Kemsley, N. (2010) *An economic impact assessment for the zoo and aquarium sector*. London, UK: British and Irish Associations of Zoos and Aquariums.

Klahr, D. (2000). *Exploring science: The cognition and development of discovery processes*. Cambridge, MA: MIT Press.

Leinhardt, G., & Knutson, K. (2010). Grandparents speak: Museum conversations across the generations. *Curator, 49*(2), 235–252.

Loucks-Horsley, S., Kapitan, R., Carlson, M. O., Kuerbis, P. J., Clark, R. C., Melle, G. M., Sachse, T. P., & Walton, E. (1990). *Elementary school science for the '90s*. Andover, MA: The NETWORK, Inc.

Lucas, A. M., Linke, R. D., & Sedgwick, P. P. (1979). Schoolchildren's criteria for 'alive'. A content analysis approach. *Journal of Psychology, 103*, 103–111.

McLaughlin, E., Smith, W. S., & Tunnicliffe, S. D. (1998). Effect on primary level students of in service teacher: Education in an informal science setting. *Journal of Science Teaching, 9*(2), 123–142.

Mithen, S. (1996). *The prehistory of the mind*. London: Thames & Hudson.

Osborne, J., Wadsworth, P., & Black, P. (1992). *Processes of life*. Liverpool, UK: University Press.

Packer, J., Ballantyne, J., & Falk, J. (2010). Exploring the impacts of wildlife tourism on visitors' long term environmental behavior and learning. *International Zoo Educators, 46*, 12–15.

Palmer, J. A. (1993). Development of concern for the environment and formative experiences of education. *Journal of Environmental Education, 24*(3), 26–30.

Patrick, P., Mathews, C. E., Ayers, D. F., & Tunnicliffe, S. D. (2007a). Conservation and education: Prominent themes in zoo mission statements. *Journal of Environmental Education, 38*(3), 53–59.

Patrick, P., Matthews, C., & Tunnicliffe, S. (2011). Using a field trip inventory to determine if listening to elementary school students' conversations, while on a zoo field trip, enhances preservice teachers' abilities to plan zoo field trips. *International Journal of Science Education*. Available at: http://www.tandfonline.com/doi/abs/10.1080/09500693.2011.620035, 1–25, iFirst article.

Patrick, P., Matthews, C., Tunnicliffe, S., & Ayers, D. (2007b). Mission statements of AZA accredited zoos: Do they say what we think they say? *International Zoo News, 54*(2), 90–98.

Patrick, P., & Middleton, J. (2011, September). *Education as a zoo mission: An analysis of the education components of zoo websites*. Poster presented at the Association of Zoos and Aquariums, Atlanta, GA.

Piaget, J. (1929). *The Child's conception of the world*. London, UK: Routledge and Kegan Paul.

Prokop, P., Kubiatko, M., & Fanoviová, J. (2007). Why do cocks crow? Children's concepts about birds. *Research in Science Education, 36*, 393–405.

Prokop, P., & Tunnicliffe, S. D. (2008). Disgusting animals. Primary school children's attitudes and myths of bats and spiders. *Eurasia Journal of Mathematics, Science and Technology Education, 4*, 87–97.

Reiss, M. J. (2005). The importance of affect in science education. In S. Alsop (Ed.), *The affective dimensions of cognition: Studies from education in the sciences* (pp. 17–25). Dordrecht: Kluwer.

Rennie, L. J., & McClafferty, T. P. (1996). Science centres and science learning. *Studies Science Education, 27*, 53–98.

Rosch, E. H., Mervis, C. B., Gray, W. D., Johnson, D. M., & Boyes-Braem, P. (1976). Basic objects in natural categories. *Cognitive Psychology, 8*(3), 382–439.

Street, B. (2010). The role of zoos and aquaria in connecting children with nature. *International Zoo Educators Journal, 46*, 6–8.

Strommen, E. (1995). Children's conceptions of forests and their inhabitants. *Journal of Research in Science Teaching, 32*(7), 683–689.

Swanagan, J. (2000). Factors influencing zoo visitors' conservation attitudes and behavior. *Journal of Environmental Education, 31*(4), 26–31.

Tanner, T. (1980). Significant life experiences. A new research area in environmental education. *Journal of Environmental Education, 11*(4), 20–24.

Tomkins, S., & Tunnicliffe, S. D. (2007). Nature tables: Stimulating children's interest in natural objects. *Journal of Biological Education, 41*(4), 150–155.

Trowbridge, J., & Mintzes, J. J. (1985). Student's alternative conceptions of animals and animal classification. *School Science and Mathematics, 85*(4), 304–316.

Trowbridge, J., & Mintzes, J. J. (1988). Alternative conceptions in animal classification: A cross age study. *Journal of Research in Science Teaching, 26*(7), 547–571.

Tunnicliffe, S. D. (1994). Why do teachers visit zoos with their pupils? *International Zoo News, 41*(5), 4–13.

Tunnicliffe, S. D. (2001). The ultimate educational resource: A visit to London zoo by first-year undergraduates in biology and education. *International Zoo News, 41*(5), 4–13.

Tunnicliffe, S. D. (2010, March). First Engagement with the environment. Children are in touch with nature. *People and Science*, p. 25.

Tunnicliffe, S. D., & Reiss, M. J. (1999a). Talking about brine shrimps: Three ways of analysing pupil conversations. *Research in Science & Technological Education, 17*(2), 203–217.

Tunnicliffe, S. D., & Reiss, M. J. (1999b). Building a model of the environment: How do children see animals. *Journal of Biological Education, 33*(4), 142–148.

Scheersoi, A., & Tunnicliffe, S. D. (2009). Engaging the interest of zoo visitors as a key to biological education. *International Zoo Educators Journal, 45*, 18–20.

Tunnicliffe, S. D., & Ueckert, C. (2007). Teaching biology-the great dilemma. *Journal of Biological Education, 41*(2), 51–52.

Vadala, C. E., Bixler, R. D., & James, J. J. (2007). Childhood play and environmental interest: Panacea or snake oil? *The Journal of Environmental Education, 39*, 1–17.

Vygotsky, L. S. (1962). *Thought and language*. Cambridge, MA: MIT Press.

Wagner, K. F., Becker, D., & Fulk, R. (2011, September). Choose the ditch. Incorporating nature play into your zoo and aquarium. *Connect*, 13–15.

Wagoner, B., & Jensen, E. (2010). Science learning at the zoo: Evaluating children's developing understanding of animals and their habitats. *Psychology & Society, 3*(1), 65–76.

Watson, A. C., Nixon, C. L., Wilson, A., & Capage, L. (1999). Social interaction skills and theory of mind in young children. *Developmental Psychology, 35*, 386–391.

Weiss, I. (1997). The status of science and mathematics teaching in the United States. *National Institute for Science Education, 1*(3), 1–8.

Chapter 10
Information Educators Need to Know About Zoo Field Trips (Useful Field Trip Information)

Museum visitors must somehow perceive information before they can store it in memory. Under normal conditions, people pay attention to things that interest them. Their interests are determined by experiences, knowledge, and feelings. This is a classic feedback loop: People learn best those things that they already know about and interest them, and people are interested in those things they learn best. (Falk & Dierking, 1992, p. 100)

Previous chapters have considered the history of zoos, visitors' knowledge, and the importance of understanding zoo visitors' discourse. This chapter builds on those chapters and explains the acquisition of appropriate knowledge and skills that visitors need in order to maximize the pedagogical and sociocultural benefits of a zoo visit. Zoo visits can be important in developing an understanding of science and the aspect of cognitive development in science literacy such as zoology, ecology, and biological conservation. An informal zoo visit has a number of functions such as formal education, recreation, entertainment, and/or social interactions. In the case of formal school visits, the teacher (organizer) organizes a field trip with learning objectives that may be different from the objectives of the students (participants). Teachers, who plan educational zoo visits, usually do so at one of three educational phases and each phase has its own pedagogical reasons: (1) Teachers plan zoo visits as an introduction to a unit or a new phenomenon hoping that the visit will stimulate students' interest in a new topic by triggering a reaction in the affective domain. (2) Teachers arrange a zoo visit in the middle of a learning unit to focus on a specific topic or a particular component of the learning scheme. (3) Teachers arrange zoo visits at the end of lesson or unit. This visit is a focused review or is used for consolidating information. The zoo visit becomes a recapitulation of knowledge or a summative assessment in which learners are required to interpret and explain aspects of animals and/or exhibits. A Teacher's visit preparation depends on the age of the children undertaking the visit, the rationale for the visit, and the familiarity of the children with the concepts being studied. However, some visits do not have a pedagogical reason. These visits are arranged as a special event for the learners and sometimes relate to the school calendar like the end of the year class outing. Many times the visits are not well planned, and classroom learning is not extended to the visit (DeWitt & Osborne, 2007).

P.G. Patrick and S.D. Tunnicliffe, *Zoo Talk*, DOI 10.1007/978-94-007-4863-7_10,
© Springer Science+Business Media Dordrecht 2013

Teachers, who spend time carefully preplanning the zoo visit, will find that they will be better able to assess the learning of the students. In the classroom, teachers would never dream of not assessing student knowledge both formatively and summatively. The essence of helping a student learn and ensuring learning is taking place is to assess the learner before, during, and after the encounter. Many people with young children assess them without thinking by asking the children questions. This is a natural mode of formative assessment and aids the learner in scaffolding knowledge. Teachers and parents can use simple questions, consider the answer, and then decide if further questioning is needed. These questioning techniques are ways of assessing with the aim of developing an understanding of the learner's place in acquiring a concept. However, assessment is more than knowing facts. Assessment is concerned with skills, processes, and the learner's place in the scaffolding of new learning. By listening to the learner, the teacher can evaluate what the students do not understand and can execute effective formative assessment. Armed with an understanding of the students' prior knowledge and understanding, the teachers can plan successful actives, experiences, and discussions that will push students to be engaged and develop new concepts. Primarily, listening to children in groups can inform a teacher about a child's progress, aid the teacher in identifying peer support, and allow the teacher to distinguish the connections students make between their prior knowledge of animals and what the students see during the visit.

Analyzing Discourse

The core of science is observation. A person's expressed model may be accessed through drawings, modeling, or talk such as interviews or general discourse (see Chap. 4). The expressed model is our only means of determining the participants' understandings. During a zoo visit, much discourse takes place between student–student, adult–student, and adult–adult.

If zoo and classroom educators are to compare the discourse that takes place in the zoo, then they should use a standard way of capturing comments and conversations. Teachers, who utilize a reflective practice of assessing their teaching, can assess student understanding during the visit by capturing the content of children's conversations. Therefore, capturing the content of what children say can be achieved by listening to and digitally recording the dialogue, analyzing the comments, and assessing the purposes of the dialogue. A standard way of capturing parents', students', teachers', and chaperones' comments and conversations will make this daunting task quicker and easier. Classroom and zoo educators can digitally record the dialogue then transcribe, read, and code the verbal interactions. However, a quicker method, which provides an overview of the conversations' content, is to use a Tunnicliffe Conversation Observation Record (TCOR) sheet (Table 10.1). Teachers can practice by recording a conversation in class, while

Table 10.1 Example of a Tunnicliffe Conversation Observation Record

Number of people in group C: # of children; A: # of adults	C: A:	C: A:	C: A:	C: A:	C: A:	Total
Animal being observed						
Management						
Example: Look. Come here. Stop that. Go there.						
Social						
Non-zoo-related talk. Example: "hello," gossip about family, group plans, watching TV shows, etc.						
Exhibit						
Anything group members say that have to do with the exhibit. Example: exhibit information, exhibit design						
Location						
This is where the animal is. Example: It is over there. Where is it? It is not here						
Informational about Animal						
Asking questions about the organisms. Adult or student providing information about the organisms						
Emotional/Affective						
Example: I like/love them. It smells. I hate them. I am afraid of them. Noises such as Ah, Oh. Is it OK? I like its hairstyle. It is so cute						
Habitat						
Example: These live in Africa						
Conservation						
Example: They are endangered. I do not think there are many of these left in the world						
Anatomy						
Example: It has a tail. It has big ears. It has 4 legs. It is brown. Look at its feet						
Behavior						
What the organism is doing. Example: It is using the bathroom. It is running. It is looking at me. It is eating. It is scratching						
Naming						
Any reference as to what to call the organism						
Other						

learners are involved in class discussions about animals. This will provide the teacher with practice and allow them to become familiar with the technique and categories prior to the zoo visit.

The TCOR is an easy to use checklist that allows the listener to quickly identify the content of conversations. On the first row, the evaluator may record the number of people in the group or they may only record the number of people who speak. The second row is a place to record the animal that is being viewed. If the TCOR is being used by a teacher, this may change as the teacher-led group moves around the zoo. If the TCOR is being used by an observer or zoo educator, the animal may stay the same, but the groups being recorded will change. The topics down the side of the TCOR are the main categories of discourse based on prior research (see Chaps. 7 and 8). The topics (main categories) may be adapted to include topics that are relevant to the user. For example, the TCOR in 9.1 may be used by zoo educators, but the topics may change to address other concepts such as animal welfare and comments that relate to the science curriculum. As the researcher listens to the conversations within a group, the researcher should make a tick mark as the topics are included in the conversation. One tick mark should be made for a conversation within an exhibit unless the researcher wants to know how many times the topic was mentioned within a conversation at an exhibit. A conversation is defined as the discourse that occurs between group members and begins when they enter the exhibit and ends when the group exits the exhibit. The following is an example of a conversation that took place among a school group (8-year-olds) in a snowy owl exhibit. The conversation shows the topics that should be ticked on the TCOR. The topics are only ticked once, because we are trying to determine the topics in each conversation not the number of times the topic was mentioned.

Snowy Owl

Boy 1:	They're like camouflaged like the bushes. [Topic: informational]
Boy 2:	Yeah!
Girl 1:	There's one up there. [Topic: location]
Boy 2:	They're sleeping. [Topic: behavior]
	The children are searching and pointing.
Girl 2:	There's the snowy owl. [Topic: naming]
Girl 2:	Ah, look! [Topic: emotional/affective and topic: management]
Girl 3:	Ah, look!
Girl 4:	Ah, look!
Girl 1:	Isn't it sweet?
Boy 1:	Oh!
Boy 1:	It's a snowy owl
Boy 2:	Yeah. I know.
Girl 3:	Ah, look! (Laughs) its ...
Boy 1:	What's it like then?
Boy 2:	Yes.
Girl 2:	It's a snowy owl!
Boy 2:	Yes. I know.
	The children are reading the signs on the cages.
Boy 2:	It's twenty years old.
Girl 1:	(Giggles)
Girl 2:	They ain't asleep.
Teacher:	Where does it say that Mike?
Boy 2:	Here.
Teacher:	It's quite old then, isn't it? It is the father of 56 chicks.
Girl 1:	What!

Boy 1:   It's had its head chopped off. (There are white rats dead on the floor of some of the cages.)

Boy 2:   (Laughs)

Boy 1:   Ugh!

Boy 2:   Sick!

Girl 1:   Look! Look!

Girl 2:   Mike look!

Girl 3:   Ugh!

Girl 4:   (Screams)

Boy 1:   It's coming out.

Boy 2:   No that's the white mice again. Oh man!

Teacher: Dan it's there (pointing to and reading label called ZOO DIET). In the zoo the owls are fed on fresh killed rats or mice.

Boy 2:   It's a zoo diet!

Girl 1:   Yuck!

Boy 2:   Owls are on a diet?

Girl 1:   I thought apples were on a diet.

Teacher: Apples. Yum, yum!

Boy 2:   I don't like green apples.

Therefore, this conversation is scored: 1 management, 1 location, 1 emotional/affective, 1 informational, 1 behavior, and 1 naming. This means that the group did not use exhibit, habitat, conservation, or anatomy comments during their time in the snowy owl exhibit.

The following conversation among a school group (7-year-olds) takes place at a panda exhibit:

Panda

Boy 1:   Hurry! Hurry up! I want to see the pandas. [Topic: management and topic: naming]

Boy 2:   Let's hurry. I am hungry! [Topic: social]

Boy 1:   Ah! It's chewing on bamboo. [Topic: emotional/affective]

Teacher: What food do they eat?

Girl 1:   Bamboo. [Topic: informational]

Girl 2:   I can't see.

Girl 1:   Ah! You can see his back.

Teacher: What is he eating?

Girl 1:   Bamboo.

Girl 2:   Yes. Bamboo.

Girl 1:   Oh! Look!

Teacher: How many shoots is he eating?

Girl 2:   Why are those things on the wall for? [Topic: exhibit]

Teacher: Its bamboo to make him feel more at home.

Boy 1:   If they made it real bamboo he'd eat it.

Teacher: That's right. They make him feel at home. Notice what he does with the sticks. He doesn't want them.

Boy 2:	He can go in there. (Boy points to the panda's entrance/exit to the exhibit.)
Teacher:	That's right he can go in there.
Girl 2:	Ah, look!
Girl 1:	He's going to sleep. He's had enough dinner. Ah! [Topic: behavior]
Girl 2:	He's cute.
Girl 1:	There are not many other pandas?
Teacher:	There are two other pandas.
Girl 1:	No. There are not many pandas anywhere.
Teacher:	You are right. There are not many pandas.
Boy 2:	They have big teeth. [Topic: anatomy]

This conversation is scored: 1 management, 1 social, 1 exhibit, 1 emotional/affective, 1 conservation, 1 informational, 1 anatomy, 1 behavior, and 1 naming. This means that the group did not use location comments during their time in the panda exhibit. This data will aid educators in determining the topics that visitors discuss in specific exhibits. Zoo educators may use this information to develop educational programs that address the topics that may not naturally occur during a visit.

Panda

Boy 1:	Hurry! Hurry up! I want to see the pandas. [Topic: management and topic: naming]
Boy 2:	Let's hurry. I am hungry! [Topic: social]
Boy 1:	Ah! It's chewing on bamboo. [Topic: emotional/affective]
Teacher:	What food do they eat?
Girl 1:	Bamboo. [Topic: informational]
Girl 2:	I can't see.
Girl 1:	Ah! You can see his back.
Teacher:	What is he eating?
Girl 1:	Bamboo.
Girl 2:	Yes. Bamboo.
Girl 1:	Oh! Look!
Teacher:	How many shoots is he eating?
Girl 2:	Why are those things on the wall for? [Topic: exhibit]
Teacher:	Its bamboo to make him feel more at home.
Boy 1:	If they made it real bamboo he'd eat it.
Teacher:	That's right. They make him feel at home. Notice what he does with the sticks. He doesn't want them.
Boy 2:	He can go in there. (Boy points to the panda's entrance/exit to the exhibit.)
Teacher:	That's right he can go in there.
Girl 2:	Ah, look!
Girl 1:	He's going to sleep. He's had enough dinner. Ah! [Topic: behavior]
Girl 2:	He's cute.
Girl 1:	There are not many other pandas?
Teacher:	There are two other pandas.

Girl 1:   No. There are not many pandas anywhere.
Teacher: You are right. There are not many pandas.
Boy 2:   They have big teeth. [Topic: anatomy]

This conversation is scored: 1 management, 1 social, 1 exhibit, 1 emotional/affective, 1 conservation, 1 informational, 1 anatomy, 1 behavior, and 1 naming. This means that the group did not use location comments during their time in the panda exhibit. This data will aid educators in determining the topics that visitors discuss in specific exhibits. Zoo educators may use this information to develop educational programs that address the topics that may not naturally occur during a visit.

The emphasis of science education has changed from learning scientific facts and information to acquiring science processing skills, developing the abilities of inquiry, and seeking evidence through the evaluation of evidence. Science teaching now highlights the learning of factual information within the context of inquiry by using relevant technologies, instructional strategies, abilities, and concepts (see www.project2061.org). Moreover, science is now taught as it relates to the socioeconomic perspective. Because of this change in focus, zoos are taking heed and are designing programs that provide schools with opportunities to select interpretative options. Schools also play a part in how the zoo visit influences how learners acquire an inquiry approach. The work done in school before the zoo visit is critical in developing an inquiry approach.

Exhibit Learning Cycle

The formal education world promotes a variety of learning cycles. Kolb's (1984) experiential learning cycle suggests a four-stage cycle consisting of experience, reflection, abstraction, and experimentation. US educators frequently use the 5 E's. The 5 E's consist of five distinct stages defined as engage, explore, explain, elaborate, and evaluate (Bybee, 1989). Since the development of the 5 E's, two of the stages have been extended to form a 7 E's learning cycle. Engage has been subdivided into elicit and engage. Evaluate has become evaluate and extend. However, the informal education world touts a more basic 3 I's exhibit learning cycle (Fig. 10.1). The 3 I's exhibit learning cycle consists of Interaction, Introduction, and Interpretation and occur during the visit. The 3 I's exhibit learning cycle encompasses the formal educational learning cycles, and the teacher can place the various stages of a formal learning cycle within the 3 I's.

The Interaction stage occurs as the visitors enter the exhibit. During this stage, visitors are becoming familiar with the exhibit by using their senses and referring to the exhibit labels. For example, visitors may be reading or viewing signage to determine the organism on display. If students are prepared for the visit, the students will have prior knowledge about the exhibit and its contents. This stage continues into the Introduction stage. The Introduction stage takes place when visitors name organisms and classify its features. In families and occasionally school groups, an

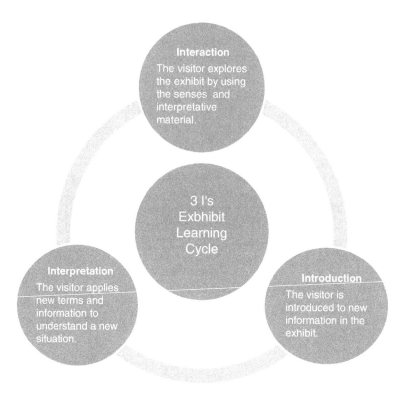

Fig. 10.1   3 I's exhibit learning cycle

adult assumes the role of information source and provides declarative discourse in which they introduce information (identifying). Teachers should prep students prior to the zoo visit by introducing new concepts. An introduction to new concepts will allow for more productive 3 I's stages. For example, children may visit the zoo with prior knowledge that an African lion has 4 legs, a tail, and sharp teeth and has characteristics similar to a domestic cat. However, they may not understand that an African lion is a vertebrate, a carnivore, and a member of the cat subgroup (taxonomy) until they visit the zoo and observe or collect this data. During a school field trip, this information may be found with the aid of a teacher or chaperone or instructional material. Students begin to construct hypotheses about the new information. The Interpretation stage happens when visitors apply their hypotheses and label information to previous exhibit information and prior knowledge. During the Interpretation stage, children apply their newly constructed ideas and test their hypotheses in a new situation or exhibit. For example, students may determine that the African lion is a cat and a vertebrate, and when the students visit other cats such as the tiger and jaguar, the students are able to associate these animals with cats and vertebrates. The Interpretation stage's discourse can be determined when students ask questions that are related to prior exhibits or knowledge.

The discourse that has been captured in zoos indicates that the conversations generated during a school visit do not contain content within the interpretation stage. The majority of conversations occur during the interaction and introduction stage. Students are involved in generating commentary about their observations, but the students are not constructing hypotheses and making predictions based on their observations. Children are not "talking science" or discussing zoological or biological science. Instead, students are "talking everyday" while identifying information with which they are familiar. Students rarely ask science questions, and when the students do ask these questions, adults rarely answer them (see Chaps. 7 and 8). The 3 I's exhibit learning cycle theory provides insight into how learning theory might assist in developing students' ability to "talk science." Educators need to increase the time students spend in the interpretation stage of the 3 I's learning cycle.

The following is an example of discourse during the Interaction stage:

Giraffe

Boy 1:   Look it's a giraffe. I spy one! It's big!
Girl 1:   Where? Where? Where?
Boy 1:   Don't you smell it? It smells. Whew. It smells.
Girl 2:   I see it. It smells. It's big.
Teacher: Is it bigger than you thought?
Boy 1:   Yes.
Boy 2:   I saw one last time.
Teacher: What else is there?
Boy 1:   A zebra!
Boy2:   Zee bra! Listen! People call them Zee bra!

The following is the same conversation as it moves into the Introduction stage:

Teacher: Did you see its tongue? Do you see how long it is? [teacher points to the sign]
Boy 1:   46 to 50 cm (18–20 inches).
Teacher: What color is its tongue?
Boy 1:   Blue-black.
Boy 2:   It's got a black tongue!
Girl 1:   There's the baby one.

The conversation now continues into the Interpretation stage:

Teacher: What else can it do?
Boy 2:   It blinks to keep the dust out.
Teacher: What else have we seen that is black and white? [referring to the zebra]
Girl 2:   Panda! Panda!
Teacher: That's right.
Teacher: What other animal have we seen that looks like the giraffe?
Boy 2:   Okapi! Okapi! Okapi!
Teacher: Yeah. You're right!

The following are examples of discourse in which the teacher and a farmer are guiding students to become more involved in learning. The discourse takes place at a farm instead of at a zoo:

Sheep (Students were asked to measure the girth of a plastic sheep that is on display.)

Boy:	You know the girth? Well it was 71 and I got 71.
Teacher:	71 what? What unit of measurement were you using?

Cows

Boy:	They're like dogs.
Farmer:	These are very expensive about £800. We're going to breed from them.
Boy:	Oh! I'll just get my dad to get me one for Christmas! [children are laughing]
Teacher:	Do you know what it means when he says breed them?

Sheep (The children were invited to put their hand in the fleece of a sheep.)

Boy 1:	Put your hand in it. It goes far down? It don't stop.
Farmer:	It does if you push far down you can feel the skin.
Boy 2:	Yeah! I got there.
Farmer:	How deep do you think the fleece is?
Boy 1:	It's about 5 or 10 inches?
Farmer:	Ten inches? Now look at your hand. It goes in up to your ring. Now pull your hand out. How long is that distance there?
Boy 2:	About 5.
Farmer:	In centimeters?
Boy 1:	Is it 4? 5?
Boy 2:	About an inch?
Farmer:	It's about 4 or 5 cm, isn't it?
Boy 2:	Yeah!
Farmer:	That's how thick her fleece is.
Boy 1:	What if you did it to the other one? It's be about 7, isn't it? The other one?
Farmer:	I don't know about the other one; it's probably got about the same thickness of fleece. In the summertime though after they have been sheared, it's only that thick, but in the winter time its much thicker.
Farmer:	Shall we let her go?

Chickens

Boy 1:	Do you want to hold it?
Boy 2:	Now I'm not touching it.
Boy 1:	Don't hold it too tight!
Girl:	How heavy do you think it is?
Boy 3:	Chicken soup!
Boy 2:	Ugh!! Ugh!!
Girl:	How many kilos?

Boy 1:	50 kilos.
Girl:	50 of them! Errr!
Boy 1:	No 50 grams!
Boy 2:	No that's about 5, 7.
Boy 1:	Well you hold it then.
Boy 2:	No I'm not touching it! I don't like animals except for dogs.
Boy 2:	I love dogs! I love cats!
Boy 1:	I got a dog!
Boy 2:	Cock-a-doodle-doo!
Boy 3:	This is going to be chopped up soon!
Boy 2:	For a chicken.
Boy 2:	I don't know what they breed them for . . .
Boy 2:	Er! Watch its neck (he jogged the bird back and forwards so it's neck lagged). Watch its neck!
Boy 1:	Wow!
Boy 3:	. . . if they're just going to chop them up....
Boy 2:	The way it looks up. Watch the way it looks up!
Boy 1:	They can breed that. They can breed that and make them bigger and make more money.
Boy 2:	Watch its head, watch its head!
Girl:	It's side to side.
Farmer:	Remember you can do this in grams and kilograms.
Boy 2:	Yes.
Farmer:	This one may not be many kilograms.
Boy 1:	Look what's it's done on the floor!
Teacher:	Why don't you try lifting the bird up to feel its weight?
Boy 1:	That's what I've been doing.
Teacher:	Remember a kilogram is about a bag of sugar. That'll give you a rough guide okay.
Boy 2:	Er! I don't want to hold it.
Boy: 2:	Give us it here.
Girl:	Here.
Girl:	You're hurting it!
Boy 2:	Ah! It's hurting my hand.
Boy 1:	Drop it then!
Teacher:	Have you got your results? Have you written down your estimate? Now we're going to do the weight.
Teacher:	Put it into the basket gently. Right children! Can you see the weighing scales? What's its weight? Just look up? Hold the bird in there. One kilo, isn't it? Can you see that? One kilo? What is it then? We're actually holding it up like that. Let it go. Gently!

Guinea Pigs

Teacher:	Don't forget to do your estimates. Girth, length, and weight.
Boy 1:	I've done my estimating.
Teacher:	Good. Now make sure you write it down before you get the actual.

Girl 1:	What do you think its estimate is?
Girl 2:	About a kilo?
Girl 1:	Most guinea pigs are rubbish! I don't like them!
Boy 1:	Shall I start measuring it?
Boy 2:	Be careful they scratch your hand.
Boy 1:	What do you measure it from? Measure its length . . . Where it comes . . .
Boy 2:	Well look, that's its shoulder isn't it?
Girl 2:	This is the best one to measure.
Boy 2:	This one scratches. I nearly cut myself. I am going to get tetanus.
Boy 1:	No, no, no! You're doing it wrong. You're doing it wrong. Look! Look!
Girl 1:	How are you going to do the girth?
Girl 2:	The girth is easy.
Teacher:	With his mouth full of apple he won't be able to bite you. Don't forget the difference between length and girth.
Boy 1:	It was 12. Put down 12.
Boy 2:	23.
Girl 1:	24.
Boy 2:	Girth say 23.
Boy 1:	Oh look at that! 23 cm.
Teacher:	Did you measure your guinea pig?
Boy 2:	Yes. Two measurements and the last one was a real one.

Language is an essential tool of science in the classroom and on field trips. Therefore, prior to the zoo visit, the science content that will be addressed during the visit should be practiced and become a part of the students' everyday agenda. The zoo visits should reinforce learning in the classroom and should be a reflection of the discourse that occurs in the classroom. Because the discourse that occurs in the zoo should be a reflection of the classroom discourse, teachers and other adults must pay attention to the form and function of the language that is used during the zoo visit. An awareness of the conversation's influence during the zoo visit should mold the form and function of the conversations that occur among students and the accompanying adults. The only way this will occur is if the children and accompanying adults are familiar with the critical attributes and behavior that are used to group organisms. Additionally, children should also understand the ways of science inquiry and how to collect data and voice questions. Likewise, adults need to familiarize themselves with the content of dialogue and how to maximize children's learning opportunities.

During the zoo visit, the focus should be on the content of what students learn and the meaning learners construct from the experience (Driver, Squires, Rushworth, & Wood-Robinson, 1994; Hein, 1995). If zoo visitors are to achieve meaningful learning, understanding how visitors construct meaning is of great importance to zoo and classroom educators. Zoos need to know what "it is" that the visitor pays attention to so that successful interpretation can be implemented.

Educators have tasks for students to complete that they hope will contribute to the students' cognitive and social learning. However, the educators and the students do not have similar expectations (Anderson, Piscitelli, & Everrett, 2008). These miscommunications or disconnects may occur when students expect to see a particular animal such as a shark, a giant panda, or a polar bear, but the zoo does not have that particular animal. When children think of the zoo, they think of the "Noah's ark megafauna," which may cause them to have unreal expectations for the visit. This unreal expectation may be thwarted by making sure that students know what they will and will not see during the visit. When students anticipate seeing large animals, but the zoo does not have megafauna, this can lead to students' disappointment and disengagement. Educators may explore what students expect from the zoo visit by asking older students to construct a mind map or a concept map (Novak, 1991) and asking younger students to draw pictures of animals. Clarification of the agendas of the zoo educators, teachers, students, and chaperones prior to the zoo visit may contribute to more effective learning, positive outcomes, and enjoyment.

When visitors approach an enclosure and enter the interaction stage, their exhibit conversation begins. In nearly two-thirds of the conversations, visitors attempt to locate the organism and comment about its location. Once the visitor sees the animal, the visitor almost always identifies the animal in some way. During the introduction stage, visitors, especially children, name the animal and have an expectation that the animal will do something. Thus, children and adults will call to the animal, knock on the glass, and make loud noises, if the animal is not active. Moreover, visitors have expectations that the animal's external anatomy will conform to their mental model of the animal. Visitors will only mention the external anatomy that does not conform to their personal mental model of the animal. During the introduction stage, visitors will mention familiar and unfamiliar external anatomy and the animal's behavior and may relate the animal's anatomy or behavior to their own or to an acquaintance's anatomy or behavior. The following are examples of students' comments in which they compare an animal's anatomy to themselves or someone they know:

Chimpanzee
Boy: I like their teeth they have the same teeth as us.

Panda
Boy: Look at the size of that one. It's big, ain't it? Look at that. That's bigger than my Dad.

Gorilla
Girl: Their hands are like ours.
Boy: If you look closely they haven't got hair on their palms or fingers. Have they?

In these examples, students are not formulating hypothesis or comparing anatomy to previously identified animals. Therefore, educators must consider how they can generate conversations during the interpretation stage that involve constructing hypotheses, making predictions, and observing.

Increasing Communication During the Interpretation Stage

To increase the possibility that the Interpretation stage may occur and be a successful learning opportunity, visitors must learn to listen, think, and communicate effectively. Therefore, teachers should manage their teaching environment so that children practice this type of discourse and become familiar with learning conversations. Providing students opportunities to practice learning talk throughout their preparatory work will allow them to use learning talk during the zoo visit and learn about biological science in a meaningful way. Learning talk in the classroom and during the Interpretation stage should aid students in (1) identifying focus questions, (2) interpreting the focus question for the group, (3) internalizing the tasks, (4) owning the investigation, (5) talking with group members to enhance their self-esteem, (6) making meaningful contributions to the discourse, (7) verbalizing thoughts, and (8) owning the experience. When students use meaningful learning talk, educators will be able to assess the student's language proficiency, listening skills, appropriate use of vocabulary, self-esteem, social skills, knowledge and understanding of facts and the science process, and enthusiasm.

However, learning talk will not be successful if students are not familiar with the vocabulary necessary to interact in discourse. Therefore, if a teacher increases students' zoo-related vocabulary, then students should be able to construct conversations that reflect their knowledge and their ability to relate prior knowledge to what they see. Vocabulary may include colors, patterns, shapes, textures, sizes, and animal names.

Asking effective questions is a skill. The questions educators and other adults ask during a zoo field trip may shape the elating situation of a field trip into an opportunity for science learning dialogue. Planning the questions that will be asked is an important part of the preparation. Teachers use questions as a formative assessment tool to check student knowledge. However, science education should use inquiry questioning as a cognitive tool that promotes thinking in learners (Chin, 2007). Questioning techniques should include inquiry questions that involve the students from the beginning of the investigation, attract the children to learning, and focus on data collection or thinking skills. For example, ask the children about their observations in the local community prior to the visit. How many mammals can you see on the school grounds? Why do you say this animal is a mammal (bird, reptile, amphibian, etc.)? What features does that animal have that belong to all mammals (birds, reptiles, amphibians, etc.)? Do all insects have wings? What is an insect? Students can practice describing and justifying the data they have extrapolated from their observations. The questions may be designed to encourage the children to obtain data from secondary sources such as an invited speaker, reference library, database, or internet search engine. If educators prepare questions and provide students with practice identifying everyday animals from still pictures and moving images, students will be more likely to observe animal anatomy and deduce new information during their zoo observations. To reach the analysis, application, and evaluation levels of Bloom's Taxonomy, students must be asked thought-provoking questions. The following are examples of questions that may be asked in zoos that might provoke interpretive discourse:

- How do we know that this is a . . . ?
- What is same and different about . . . ?
- Why do ——, ——, and —— look similar?
- How would you explain the . . . ?
- Where do you think that animal lives in the wild? Why?
- What do the animal's characteristics tell us about the animal and what it eats?
- How are all these animals the same?
- Why is —— true?

Moreover, when educators want to trigger interpretive discourse, different types of questions will be helpful in guiding the Interpretation stage. Data processing questions should be used after the children have observed the organisms and collected data. A data processing question may give students a boost in confidence and allow them to interpret their results. Students who have been asked to solve a challenge have been heard saying, "I think we should do this . . ." and then continuing to justify the suggestion. Closure questions may be used at the end of the exhibit observation. These questions focus on a topic trigger. We want the children to think of what they have observed thus far and think about how their ideas can be summarized. For example, a student might say, "All mammals have ear flaps is what we found out." Thought-stimulating questions should be used at the beginning of the Interpretation stage. By asking a pertinent, thought-provoking question, you can stimulate a child to center his/her thoughts on the exhibit and the organism. Thought-stimulating questions encourage children to think divergently, apply knowledge, and formulate a creative response. However, specific divergent questions or extension questions are more relevant during the beginning of the Introduction stage when children are formulating their hypotheses. This type of questioning invites children to contribute more on the subject or redirects questions when the dialogue is heading off topic and needs redirecting. Conversely, convergent questions are useful when the group is summarizing the exhibit and comparing observations from other exhibits. Evaluation questions may be used throughout the 3 I's exhibit learning cycle as the educator deems them appropriate. These formative questions provide the teacher with information concerning the students' progress and understanding, lead to feedback for the students, and reinforce learning. Justification questions encourage children to say more about a statement they have made. Clarification questions invite children to put an idea into their own words. Summative questions are used at the end of the Interpretation stage and are designed to find out more about the student's understanding. Eliciting questions are usually focused, evoke a reaction, and draw out information from the child. These questions are used to get an initial reaction and may be asked during the interaction stage. For example, an educator may ask, "What do you think?" "What do you know about the panda?" "Who has seen a caterpillar?" "Has anyone ever . . . ?" However, these questions should relate to the educator's goals for the exhibit and should lead students to the discussion that will occur in the Interpretation stage. When the

Table 10.2 An example of questions that students develop prior to a zoo visit and answer during the zoo visit

Task/question/challenge	Example 1	Example 2	Example 3
Do all mammals have a prehensile tail?			
Which of the animals are brown and white?			
Which animals use camouflage?			
What animals live in the Arctic?			
Thinking questions			
Compare three animals that have prehensile tail			
Observe three animals and compare their behavior			
Describe the differences and similarities in the legs of four animals			

group becomes too involved in the many details of the exhibit and the excitement of the visit, they may lose sight of their original task. Redirecting questions can be used in all stages of the 3 I's exhibit learning cycle to bring the group back to the task at hand. Educators may ask students about what they are observing, how their observations relate to their previous knowledge, or where they have seen another organism with similar structures. We are not suggesting that all of these question types be used in every exhibit or during every visit. What we are suggesting is that by changing the type of questioning that occurs during the visit, you will change the discourse that occurs during the Interpretation stage.

To increase the likelihood that learning will occur during the 3 I's exhibit learning cycle, educators can teach and model observing with meaning. Learning to observe with meaning rather than superficial looking is a crucial skill needed to develop science skills. Young children can be encouraged to observe by making signs that they match with the zoo's animals, exhibits, or signage. Prepare children to focus on animals by reading animal stories about animals they will see during the field trip. Older children can use binoculars to look more closely at and observe animals. Moreover, children can practice using hand lenses or magnifying glasses at school and then use them at the zoo. Teachers can use zoo maps to invite the children to say which animal they would like to see. Children can begin observing animals by looking at virtual zoos. To increase the likelihood that students will observe, students can plan what they would like to learn while they are visiting the zoo. Tables 10.2 and 10.3 show examples of students planning what they want to learn during a zoo field trip. Pre-visit planning provides students with an opportunity to plan their own learning instead of the teacher directing the information that will be gathered.

As discussed in Chaps. 7 and 8, the presence of an adult within a visitor group focuses the conversations on knowledge source comments. Therefore, when children are not with a teacher-led group, they learn about the world from family members and other adults that "teach" aspects that the adult considers important. Teaching names for things is one aspect of how adults teach. There is little "talking science" that occurs between the adults and children. Children are not asked to make predictions based on observations or previous work carried out at school. Moreover, the children are not asked to further examine their observations and determine

Table 10.3 Animal characteristics that students determine they will identify during the field trip

Characteristic	Animal	Animal	Animal	Animal	Animal	Animal
Wings						
Feathers						
Teeth						
Fur						
Scales						
More than 4 legs						
Tail						

whether their prediction is correct. Visitor groups look at animal exhibits in isolation and not relationally. Visitors look at single exhibits, move on to another, and do not refer back to previous exhibits.

More Ideas

Nature Tables

Heightening awareness of local organisms can be achieved through having an observational nature table or animals table with real specimens, models, or preserved specimens (although care must be taken of the preservative used) and biofacts such as pelts, horns, teeth, and bones. Primary children can have an animal table of toy animals, which can lead to discussing of how the toys are like the animal they represent and how they are different. Sometimes such can be borrowed from the zoo in outreach packs. Other organizations may have outreach packs that can be borrowed. Frankfurt University Biology Department has complied conservation suitcases which contain information and games about aspects of conservation such as the start of a rain forest and the intercommunication of the ecosystem.

Some universities provide outreach labs, which they bring to a school. For example, the traveling zoo from the Whitney Laboratory in Florida. Docents (volunteers) take invertebrates from the rescan lab, some of which are species used in research, to schools and guide the learners as they observe them, through those animals. The aquatic zoo animals may include blue crab, clam, conch, fiddler crab, horseshoe crab, mud snail, sea cucumber, sea star (starfish in Europe), sea urchin, slipper lobster, spiny lobster, and whelk. Through firsthand observations and interaction with specimens, the aim is that learners carry out inquiry skills and learn fundamental biological facts such than an animal needs certain things in order to survive, such as be able to protect itself and reproduce for the survival of the species. The animals must also be able to obtain food, ware, and oxygen. The traveling zoo contains suggestions for students to discuss and extend their knowledge and understanding (information is available at http://www.whitney.ufl.edu/education/program-overviews.html). Labs (practical work sessions) of a similar nature are often offered by zoos and aquaria on site, for instance, in the USA, Monterey Bay run such. The aquarium in Copenhagen has a lab where schools can book to attend.

Physical Science and Hands-On Activities Pre-visit

The visitors comment upon that which they see, that which catches their interest and interpret the animals from a human view point. In a giraffe exhibit children commented "I'd only come past his knee" or "That's as tall as a house!" The visitors are not seeking to explain the animals in terms of biological science. The giraffe was recognized and allocated to an identity category; the children did not categorize it within zoological taxonomy.

The characteristics of animals and is the defining one cited by elementary children. Most animals can make some observable movements and most posses the power of locomotion—being able to move from place to place. The principle of human walking is that of a pendulum. Before discussing movement in the animals, it is useful to discuss how humans move. Try constructing and then observing a pendulum; a simple one can be made from a piece of string and blob of modeling clay around a paper clip tied to the end of the string or tied on a metal washer to act as the bob in the pendulum. The bob repeatedly rises and falls and slows down and speeds up. As the bob swings down, it loses height and thus potential energy; when it rises up, it gains height and so kinetic energy. Energy is swapped backward and forward. In waking, the person slows down and rises in the first half of each step but speeds up and falls in the second half—try it!

The total mechanical energy—potential plus kinetic—should remain constant. Leg muscles have to give a boost at one stage (increasing the mechanical energy of the body) and act like a brake at other times (converting mechanical energy to heat). The principle of running is that of a bouncing ball. In human running the Achilles tendons in the legs give bounce. Locomotory movement has many functions—food finding, escaping, hunting, and finding a mate. Single-celled organisms move by using a cilia or flagella and by amoeboid "crawling," for example. Talk about such with children and together or in small groups or even individually ask them to make a table of various types of locomotion. Many celled animals move using a number of different means: crawling—e.g., earthworms, snails; walk, run, fly, and swim—mammals; fly—birds and insects; and move through water—fish (some birds) and insects (especially larvae), e.g., caddis fly and great diving beetle.

If you are working with older students, invite the children to set up the investigation themselves designing. If you design a sheet to guide learners in their planning before a visit or in practical work in school, the sheet can be adapted for observation recording in the zoo through changing the tense; altering the tense, you can use this as a record sheet during the investigation and as a reporting sheet after the investigation. Possible responses are given in inverted commas. Animals have adapted other natural environments where they have evolved. They have adaptations from the basic pattern of their phyla which are suited to that environment. Preparation for a zoo visit can usefully include discussions about the needs of animals and how they are met in different types of environment starting with biomes and hone looking at particular niches.

Patterns of Animal Anatomy

Establish with learners the basic pattern for the animal you are expecting to observe at the zoo. These are practically all coelomates or three-layered animals. The majority of which are bilaterally symmetrical. The cnidarians, jellyfish and sea anemones, are two-layered and radically symmetrical. Ensure that children know what it is to be radically symmetrical and bilaterally symmetrical and have them identify these characters in photographs and films so they can allocate animals seen during their zoo visit to the appropriate category. The animal on exhibit may be mostly vertebrates with four appendages, head and spinal cord and post-anal tail. Appendages in vertebrates have the basic pattern of one bone, two bones then 5 sets of carpals, tarsal, and phalanges. Children again can practice allocating images of animals or real ones they see every day to a category of vertebrate, mammal, bird, amphibian, and fish. Discuss with learners the basic vertebrate pattern modified for different habitats such as water and land, land and develop the theme when in the zoo.

Invertebrates most often seen in zoos are arthropods particularly insects, with external skeletons and through guts. The insect patterns of three parts to the body (head, thorax, and abdomen) with three pairs of leg on the thorax, the arachnids with two parts to the body and four pairs of legs on the cephalothoraxes, and the annelid worms with a segmented body with a through gut are main groups.

Discuss what the needs of animals are—food, habitat, sensing danger, and locomotion are examples. Discuss how some of the animals to be observed on the zoo visit achieve this. In the case of locomotion on a surface, how is such achieved when the animal walks on sand or snow? How do animals sense danger? How do they blend into their environment? What happens to animals that live in a cold climate? What happens to mammals that live in areas with snow in winter and none in summer so they do not show up when the ground coverings is different to their coat color? How do animals where the ambient temperature is low keep warm? How do animals that live in water adapt for movement? How do mammals in water keep warm? How do fish survive freezing conditions? How do mammals living in hot places keep cool?

Hands-On Activities Pre-visit

You could ask your learners to find out how elephants for instance keep cool. What part of their body can they use to help cool their blood? Elephants use their earflaps for several things. One is fanning so you could challenge learners to see the effects of fanning on a body. For example, fill similar dishes with hot water, take the temperature of each dish and fan only one. Make sure the fanning of the one does not affect the other. Use a heat strip to take the temperature of the water in both dishes before and after fanning. Then consider other roles of elephant's ears. Introduce another activity connected with ears (more correctly earflaps) by inviting the children to listen to something, for example, a whisper, a muffled drum, and a coin dropping with (a) just ears, (b) one ear, (c) ears cupped with their hands, and (d) with "ears" made from cardboard or from paper cups. Making ears can be introduced as a design technology activity. How will the cups be kept on their heads? What materials are suitable to make ears from? How will they shape and join the materials? If you have data logging equipment, the effect of an "ear" on receiving sound generated by a standard item such as a buzzer and of no ear can be monitored and recorded. Children can predict and experiment on the effect of the size of an "ear"! For younger children, set up the investigations for them using different sounds for each way of hearing and ask the children if it is a fair test.

Reproduction is one of the characteristics of living things; in animals, this is usually of a sexual nature requiring a male and a female. How are various species adapted to attract a mate of the opposite sex? Which of the animals in the area to be visited display sexual dimorphism where the members of the two sexes look very different from each other although still "built" on the same basic plan? Peacock and peahens and lion and lionesses are classic examples of sexual dimorphism. Study the website of the zoo to be visited and find which animals have this dimorphism in a pronounced way and which have much slighter variation.

An interesting preparatory activity especially for younger children who relate animals to them is to compare a function found in other animals and then how humans affect this and how a named animal achieves the function. For example, consider a set of questions as follows:

Giraffe: A suggested question sheet to guide discussion and interaction activities.
What group of animals does a giraffe belong? Find out where giraffes live.
What do they eat? What is their size compared to you? How many times taller than you are they?
Their neck is much longer than ours, so how many neck bones do they have? Do we have?
Is it the same or different from us human beings? What is the same? What is different?
Do you see what giraffes see? Look at a friend standing straight in front of you.
What do you see straight in front of you? What would you see if you were a giraffe and looked at your friend in front of you?

Draw what you see and what you think a giraffe would see looking straight ahead at
   your friend.
Make a periscope so you can see as a giraffe does.

Periscope Directions: You need a long tube and two mirrors (unbreakable), scissors,
tape or strings, and glue. Work out where to make a slot at the top of the tube and
at the bottom so you can place the one of the two flat mirrors at either end of the
tube so that what is seen in the top mirror reflects to the bottom one and that reflects
into your eye. This sequence needs working out by holding the mirrors for each
other until the alignment is such to obtain the effect wanted. How do you obtain
your food? How do you put it to your mouth? How does a giraffe find its food?
From where does it obtain its food? How does it take the food from its source to its
mouth? How are they fed in a zoo?

   How often do you drink water? From where do you obtain the water? From where
do giraffes obtain water? How do they reach it? Make a model from modeling clay
and headless match sticks to represent a giraffe standing up, then alter the model so
that the giraffe drinks. What do you have to do? Why?

Zoo Kits

The following set of items is particularly useful with 3–7-year-olds:

Aim: To focus on observations of animals in the zoo
Objectives: Children will observe the color, pattern, shape, and size of animals,
   as well as developing their knowledge of animal kinds and associated language
   skills.

A cardboard eye spy tube (made from kitchen towel or other ready-made card tubes
or such) can be constructed using stiff card and glue and tying the tube with string
or taping it with string tape.

Materials:
A magnifying sheet or hand lens
A set of color cards
A set of pattern cards
A farm animal
A zoo animal
A touch, feel, and talk card

   Animals particularly useful to observe when using such an observation kit are
the flamingos, lions, tigers, penguins, elephants, farm animals, reptiles for some
colors and textures, zebra, camels, and giraffe. Plan your route through the zoo to by
marking the route on a zoo map and provide copies to each student and chaperone.
Have students practice reading the map prior to the zoo visit. Before you visit,
it is necessary to make preparation with orientation activities and plan follow-up
activities as well as construct your observation tools. These can usefully be divided
into four categories:

1. Vocabulary

   Develop the relevant vocabulary for color, pattern, shape, and size. Children should know black, white, stripped, spotted, rings, plain, patterned, gray, brown, red, green, blue, yellow, and pink. Practice describing plain and patterned things in the classroom using color words.

   Develop the vocabulary for textures—hard, soft, dull, shiny, warm, cold, rough, smooth, and soft. Practice describing familiar items in the classroom using these descriptive words.

2. Variety of life

   Introduce the children to the shapes and names of some of the animals they will see during their visit, especially the ones noted in the suggested route. Use toy animals, templates and jigsaws, and books, as well as slides and videos to introduce the animals.

3. Observation games

   Try playing I SPY in the classroom: I SPY colors, I SPY shapes, and I SPY patterns. Use your hand clap as a time measure. "How many circles can you find between my hand claps?" Play matching words to patterns, words to shapes.

4. Firsthand observations

Practice with the children using a hand lens or so that they know what it does. Look at news for print practice. Children also enjoy using the magnifying glass to look at their skin. Practice recording behaviors observed.

Thus, a zoo visit can reinforce and apply the science inquiry process—observing, raising questions, planning, auctioning, recording, evaluation, working as a team, and reporting.

Besides taxonomy and behavior, other aspects of biological science can be studied such as:

DNA Fingerprinting

Many zoos will provide talks for older learners about DNA fingerprinting and associated topics, which skills are the scientific research that is occurring in many zoos today together with their work in situ with particular species. Taxonomists classify organisms in a way that reflects their biological ancestry. Because the ancestral relationships are complex, the taxonomic schemes are also complex. The DNA sequence of certain gene regions on chromosomes can be used to identify species and establish the closeness of relationships (as in human paternity cases) and hence the evolution of species. The technique is also used for forensic purpose such as tracing the origins of illegally imported bush meat, which is often brought in passenger's suitcases (Ogden, 2009). Zoos explain such conservation and biological research initiatives on their websites. Zoo education departments are usually delighted to provide such information. Further information is also available on natural history museums website such as American Museum of Natural History in New York.

Teachers need to develop their own pre-visit (and post-visit) activities that are relevant to the curriculum. If the zoo field trip is to be successful, the adults accompanying the children must be adequately briefed. The visit experience for children can differ depending on which adult is managing the groups. Advice for organizing field trips can be obtained from the zoo's education department and from various publications (e.g., Braund and Reiss, 2004; Melber, 2008). The success of zoo visits depends on the aims and objectives for the visit set by the organizers. Within the sociocultural interactions and learning processes, individuals make sense of the world and cognitive development occurs (Bandura, 1986, 2001). Learning is embedded in the social interactions and the cultural dynamics of the group. Therefore, the verbal interactions between students, teachers, and/or other adults are an important aspect of the field trip experience. However, effective preparatory work can help the visits achieve these objectives.

To encourage students to reflect on previous exhibits and prior knowledge, teachers must develop pre-visit, during-visit, and post-visit activities that are relevant to the curriculum. Moreover, the adults accompanying the children must be adequately briefed prior to the visit. In Chap. 8 we discussed differences in family and group conversations. In the following chapter we provide a theoretical framework for the importance of organizing field trips. The success of a zoo visit depends on the aims and objectives developed by the visit organizer and effective preparatory work can prepare the students for achieving these objectives. Therefore, field trip developers should be aware of the need for preparation and follow-up activities that are connected to a zoo visit.

References

Anderson, A., Piscitelli, B., & Everrett, M. (2008). Competing agendas: Young children's museum field trips. *Curator, 51*(3), 253–273.

Bandura, A. (1986). *Social foundations of thought and action: A social cognitive theory.* Englewood Cliffs, NJ: Prentice-Hall.

Bandura, A. (2001). Social cognitive theory. *Annual Review of Psychology, 52*(1), 1–26.

Braund, M., & Reiss, M. (2004). *Learning science outside the classroom.* London: Routledge & Falmer.

Bybee, R. W. (1989). *Science and technology education for the elementary years: Frameworks for curriculum and instruction.* Washington, DC: The National Center for Improving Instruction.

Chin, C. (2007). Teacher questioning in science classrooms: Approaches that stimulate productive thinking. *Journal of Research in Science Teaching, 44*(6), 815–843.

DeWitt, J., & Osborne, J. F. (2007). Supporting teachers on science-focused school trips: Towards an integrated framework of theory and practice. *International Journal of Science Education, 29*(10), 685–710.

Driver, R., Squires, A., Rushworth, P., & Wood-Robinson, V. (1994). *Making sense of secondary science: Research into children's ideas.* London, UK: Falmer Press.

Falk, J. H. (2009). *Identity and the museum visitor experience.* Walnut Creek, CA: Left Coast Press.

Falk, J. H., & Dierking, L. (1992). *The museum experience.* Washington, DC: Whalesback Books.

Hein, G. E. (1995). Evaluating teaching and learning in museums. In E. Hooper-Greenhill (Ed.), *Museums, media, message* (pp. 189–203). London: Routledge.

Kolb, D. A. (1984). *Experiential learning: Experience as the source of learning and development.* Englewood Cliffs, NJ: Prentice-Hall.

Melber, L. (2008). *Informal learning and field trips: Engaging students in standards-based experiences across the K-5 curriculum.* Thousand Oaks, CA: Corwin.

Novak, J. D. (1991). Clarify with concept maps: A tool for students and teachers alike. *The Science Teacher, 58,* 45–49.

Ogden, R. (2009, April). The use of wildlife DNA forensic methods to investigate the illegal meat trade. Paper presented at the Zoological Society of London. UK Bushmeat Working Group Meeting.

Chapter 11
Zoo Field Trip Design

As an examination of environmental curricula will reveal, what might be seen at first glance as model approaches to incorporating environmental issues into the public school curriculum often turn out to be the most problematic. What needs to be given special attention is the ideology that is reinforced along with the scientific content of the lesson. (Bowers, 1997, p. 245)

As environmentalists begin to give close attention to what is being taught in public schools, it will become apparent that high school teachers are more concerned with the knowledge of content areas, while in the early grades the knowledge of content takes more of a backseat to promoting social skills and a positive self-concept. (Bowers, 1997, p. 247)

Observing animals in the classroom and immediately around the home or school allows students to practice making and recording observations (e.g., Patrick & Getz, 2008). However, in England fewer animals are being kept at school. This decline is thought to be due to a number of factors, one of which is the result of increasing health and safety awareness (Association of Science Education, 2001) involved in keeping classroom animals. Therefore, low-maintenance invertebrates, such as spiders, mealworms, and brine shrimp, are ideal animals for teachers to keep in school. Eighty-one percent of AZA-accredited zoos offer outreach programs which take animals into schools with a trained educator/keeper (Patrick & Tunnicliffe, 2011). Zoo outreach visits often lead to a zoo field trip (Tunnicliffe, 1993).

Even though the personal hands-on interactions students have with organisms may be limited, some teachers and parents proclaim that children do not need to visit zoos and that zoos are unnecessary. Zoo visits are unnecessary because zoo animals can be seen in the media, on television, on DVDs, and on the Internet. While these assertions are true, these interactions are impersonal, vicarious, and voyeuristic. Furthermore, the images are two-dimensional, reduced in size, and do not provoke firsthand emotive or affective interactions (e.g., smell). However, these images do provide a useful starting point for discussions about animals and habitats and offer opportunities to hone observation skills.

A museum is an organic institution or dedicated space open to all wherein a genuine artefact or a collection of genuine artefacts of aesthetic, archaeological, cultural, historical, social,

P.G. Patrick and S.D. Tunnicliffe, *Zoo Talk*, DOI 10.1007/978-94-007-4863-7_11,
© Springer Science+Business Media Dordrecht 2013

or spiritual importance and interest from any place or time is preserved, conserved and displayed in a manner in keeping with its intrinsic and endued worth. As such, it is an informative means of storing national, cultural and collective memories, where people can explore, interact, contemplate, be inspired by, learn about and enjoy their own and others' cultural heritage. (Talboys, 1996, p. 17)

Talboys' definition allows for a generous interpretation of what qualifies as a museum and acknowledges the educational experiences that may take place. Using Talboys' (1996) definition of a museum, research concerning other types of museums can be generalized to the zoo–museum experience.

A review of the informal education literature reveals extensive research related to (1) learning in museums and zoos, (2) the rationale for visiting informal institutions, and (3) field trip design. In the past 40 years, researchers have begun to focus on the importance of the field trip design (Anderson, Lucas, Ginns, & Dierking, 2000; Falk, Martin, & Balling, 1978; Henriksen & Jorde, 2001; Orion & Hofstein, 1991; Rebar, 2010; Rebar & Enochs, 2010). Krepel and Duvall (1981) define a field trip as "a trip arranged by the school and undertaken for educational purposes, in which the students go to places where the materials of instruction may be observed and studied directly in their functional setting: for example, a trip to a factory, a city waterworks, a library, a museum etc." (p. 7). Since the inception of studies in field trip design, research has focused on (1) planning field trips, (2) learning during field trips, and (3) social interactions that take place during a field trip (e.g., Anderson, 1973; Clark, 1943; Curtis, 1944; Harvey, 1951; Henriksen & Jorde, 2001; Howland, 1962; Kisiel, 2007, 2010; Krepel & Duvall, 1973; Rebar, 2010; Rebar & Enochs, 2010; Ruth, 1962).

Research has shown that field trips are exciting, memorable, social, and to some extent educational. Falk and Dierking (1997) interviewed school-aged children and adults to unveil their field trip memories. Ninety-six percent of participants were able to recall details of a field trip. "Field trips are consequential experiences in children's lives . . . [that] result in highly salient and indelible memories" (Falk & Dierking, 1997, pp. 216–217). Field trips, once considered to be a leisure day away from school, are being investigated to better understand their enormous teaching potential.

This chapter focuses on the rationale behind the zoo field trip and the components of a successful zoo field trip design. It is important to review the literature surrounding zoo field trip design in order to gain a better understanding of the developmental history of the literature, identify existing gaps in the literature, and provide potential directions for future research.

Rationale for Visiting the Zoo: Animals

Zoos are places where wild animals are kept for public display. The opportunity to see these animals and learn more about them is a major reason that teachers organize field trips specifically to a zoo. Students also view zoos as places to see animals. Two hundred and ninety-five students, ages 10–12 years, were asked to list what they knew about zoos. All 295 students named animals as the first thing they knew about zoos. Additionally, the students stated that zoos were places to learn about animals (Patrick, 2011). Wagoner and Jensen (2010) found that children's knowledge of animals and their habitats is refined by zoo education programs, and a field trip to the zoo can be an important animal learning experience. However, prior to taking a zoo field trip, teachers should understand what children know about animals, what they learn about animals during a zoo visit, and what children want to learn about animals. Understanding students' animal knowledge is an important aspect of structuring an effective, age-appropriate zoo field trip.

Kellert (1985) conducted a study to assess 267 children's attitudes toward animals and their animal knowledge. The students were from the second, 5th, 8th, and 11th grades. Several interesting findings relate to field trip design: (1) Most children had a limited knowledge of animals and a lack of ecological understanding. (2) Children were able to identify the foods animals eat and the basic biological characteristics of animals. (3) Urban children knew less about animals than do rural children. (4) Girls knew less about animals than boys. (5) Children of Color knew less about animals than Caucasian children. (6) The three main attitudes children have toward animals are humanistic, naturalistic, and negativistic. Humanistic attitudes are defined by an interest in animals such as pets. Naturalistic attitudes are an innate interest in wild animals and nature. Negativistic attitudes are characterized by an indifference, dislike, or fear of animals. (7) There are three major transitions in animal knowledge from 2nd to 11th grade. Second to fifth graders possess an emotional concern and affection for animals. Fifth to eighth grade retain factual and cognitive understandings of animals. Eighth to eleventh graders are more concerned with the ethical and ecological issues animals face.

In 2004, Myers, Saunders, and Garrett utilized 171 children's (ages 4–14) drawings and interviews to determine children's knowledge of animals' physiological needs. Similar to Kellert (1985), Myers et al. have found the following: (1) 7- to 10-year-olds often mention physiological needs. (2) Children 9 years old and older mention habitat. (3) Ecological and conservation comments increase with age. (4) Conservation is mentioned by children 10 years of age or older. (5) As children age their anthropomorphic ideas of animals evolve into realistic concepts. Even though most conversations that occur in zoo exhibits do not include conservation comments, conservation-related discourse is on the rise in zoos. During a conversation between a teacher and an 8-year-old student at an endangered black rhinoceros exhibit, the girl commented, "Most pandas are getting killed. Aren't they getting extinct?"

Rationale for Visiting the Zoo: Educational

Curriculum learning can take place inside the classroom or in informal science settings. Formal learning is planned by educators as part of the learners' curriculum, and educators plan field trips to informal science settings with definite educational objectives (Tunnicliffe, 1994; Wolins, Jensen, & Ulzheimer, 1992). When teachers decide to take students on a field trip to the zoo, teachers believe that the visit will be an educational experience. Teachers who take students on field trips are seeking out-of-classroom experiences for students that cannot be provided within the classroom (Cox-Petersen, Marsh, Kisiel, & Melber, 2003; Kisiel, 2003a, 2006a) and have found museums to be an exceptional experiential learning resource that complements and/or enriches school curriculum (Berry, 1998; Kisiel, 2006a; Sheppard, 2000). Falk and Dierking (1992) suggested that significant cognitive learning could occur on field trips to informal science settings. Field trips can be a learning experience, if teachers identify the benefits, prepare for the visit, and understand informal learning. However, the rationale behind taking a field trip to the zoo varies from teacher to teacher and, on a grander scale, varies among pedagogical and educational philosophies.

Kisiel (2005) interviewed 115 teachers to determine their motivations for organizing a science center field trip. Ninety percent of the teachers surveyed planned a field trip because it connected with the classroom curriculum, 39% planned a field trip because it would expose their students to new experiences, and 30% viewed the field trip as a general learning experience. However, fewer teachers (18%) planned a field trip to increase the interest and motivation of their students or to encourage a pattern of lifelong learning beyond the classroom (13%). This study indicates that teachers view a field trip as a leisure activity that happens to connect to the curriculum. Teachers do not see field trips as an opportunity to model learning outside of the classroom. However, Kisiel suggests that field trips provide a learning experience that differs from formal classroom learning.

Even though Pace and Tesi (2004) and Braund and Reiss (2006) do not specifically study zoo field trips, their research is of importance in the rationale for out-of-the-classroom experiences. Pace and Tesi (2004) interviewed eight adults to determine the impact of childhood field trips on long-term interests, education, and overall perceptions. They concluded from the responses that field trips to museums, zoos, and historical sites effectively reinforced the classroom curriculum; increased camaraderie between students, teachers, and chaperones; and were much anticipated. The most memorable field trip experiences were overnight trips, such as camping. Seventy-five percent of the adults viewed field trips as an opportunity to escape the everyday routine of the classroom for a novel and exciting adventure. Field trips create long-lasting memories that last into adulthood (Falk & Dierking, 1997; Pace & Tesi, 2004). Braund and Reiss (2006) suggest that the nature of learning requires practical, in-school work, and real-life, out-of-school experiences. Authentic science curriculum truly exists when both of these constructs are utilized. A well-designed zoo field trip combines the traditional classroom experiences with informal learning to benefit learners in five ways: (1) improve development and integration of concepts, (2) extend and authenticate classroom learning, (3) provide access to rare materials, (4) stimulate additional learning, and (5) promote positive social outcomes. Moreover, learning in informal contexts has been recommended as an important element in promoting interest in science, motivating student/teacher and student/student interactions, and increasing knowledge (Pedretti, 2002).

To determine if zoo visits are an educational experience, Mallapur, Waran, and Sinha (2008) asked zoo visitors and non-visitors questions concerning the conservation and natural history of lion-tailed macaques. Mallapur et al. have found that zoo visitors' knowledge of the conservation and natural history of lion-tailed macaques is significantly higher than the general public. Visitors' conservation knowledge and learning is influenced by seeing animals in replications of their natural environments, observing animal behavior, connecting with prior animal knowledge and experiences, and close encounters with wildlife (Ballantyne, Packer, Hughes, & Dierking, 2007). Moreover, visiting zoos impacts the conservation attitudes and understanding of visitors, resulting in a stronger connection to nature and an enhanced understanding of wildlife (Falk et al., 2007).

Although rare, museum, zoo, and aquarium field trips can have a cross-curricular approach (Tunnicliffe, 1992; Tunnicliffe, 1993). Cross-curricular field trips incorporate language arts, science, math, social studies, etc. The integration of language arts in science has been shown to enhance the participation of female students in science (Lockheed, 1985), and elementary teachers feel more comfortable teaching language arts than science (Weiss, 1997). Therefore, a cross-curricular approach that incorporates an informal science setting may increase the likelihood that teachers will provide the students with valuable science concepts. Research has found that reading orientation stories in language arts prior to the zoo visit does have a marked effect on the children's retention and understanding of the visit (McLaughlin, Smith, & Tunnicliffe, 1998). If a zoo plans integral, cross-curricular units, teachers may come to view the zoo as a useful teaching resource.

The rationale for taking students on zoo field trips has been established. Research shows that zoo visits and field trips are learning opportunities. Additionally, field trips in general are beneficial to student development and socialization. Zoo field trips have positive effects on students' knowledge of endangered species, expose students to novel situations and animals, and influence students' use of science vocabulary. However, a zoo field trip must include the aspects of a successful field trip design.

Learning During a Zoo Field Trip

When teachers organize a zoo field trip, their first concerns consist of signing permission slips, organizing chaperones, keeping up with lunches, and making sure that students do not get lost. However, a teacher also must consider how he or she will make the most of the educational experience and optimize students' learning. Teachers who identify field trips as educational destinations (Rosenfeld, 1980; Tunnicliffe, 1994) and take their students to the zoo for specific learning goals (Tunnicliffe, Lucas, & Osborne, 1997) should be aware of the psychological needs of visitors, the key factors of informal learning, the five stages of a zoo visit, and the characteristics of a successful informal learning experience.

Perry (1992, 1993) has identified six psychological needs of museum visitors, all of which must be met for a museum experience to be successful and educational. The six needs are (1) curiosity—the visitor wants to see new things; (2) confidence—the visitor wants to have a good time while feeling good about themselves and the experience; (3) challenge—the visitor wants to be challenged; (4) control—the visitor wants to feel like they are in charge of the visit; (5) play— the visitor wants to have fun during the visit; and 6) communication—the visitor wants to exchange information and communicate with others.

Falk and Dierking (2000) have defined three contexts that influence museum learning. In their contextual model of learning, those contexts include (1) personal context, which includes the individualized prior knowledge, interest, motivation, expectation, and experience that a visitor brings to the museum; (2) sociocultural context, which includes the possibility that learning, in an informal learning environment, may be influenced by people in the group and outside of the group; and (3) physical context, which encompasses the entire physical learning environment.

Tunnicliffe (1999) proposes five stages of a zoo visit, which are a useful framework for organizing a zoo field trip: (1) orientation, (2) concentrated or focused looking, (3) leisure looking, (4) completion, and (5) consolidation. During the orientation stage, students adjust to and find their way around the site. The orientation stage allows students to focus on the zoo experience instead of spending time adapting to the new setting. This stage can be shortened by preparing the students ahead of time with zoo maps and videos. The concentrated or focused looking stage involves the students focusing on learning tasks or activities. When students begin to lose focus and their concentration wanes, students enter the leisure looking stage. Realistically, students are not expected to be focused throughout the entire trip. However, if the field trip is not well planned and structured, the leisure looking stage can occur earlier in the visit. As the field trip comes to an end, there is a completion stage. During the completion stage, students prepare to leave the zoo and turn their attention to visiting the gift shop and going home. The consolidation stage is the teacher's responsibility. When students return to the classroom, teachers must relate the zoo visit to the formal classroom learning.

A closer look at the concentrated or focused looking stage reveals five levels (Tunnicliffe, 1999). Tunnicliffe's five levels of the concentrated or focused looking stage have been intertwined with the six levels of learning in Bloom's Taxonomy (Krathwohl, 2002) to define the Zoo Cognition Hierarchy. Tunnicliffe defines the cognitive gain in zoos, from simple conversations (level 1) to observing and interpreting data (level 5). Bloom defines the cognitive gains, from simple recall of facts to the use of knowledge. In both instances, the learning is not meant to be all in one category, but each level of representative learning is essential to the next. The comparison between Tunnicliffe's levels of looking and Bloom's Taxonomy provide a conceptual framework for future research in zoo learning. Conversations that take place within the exhibits could be coded to determine the cognitive level at which visitors are interacting. Table 11.1 depicts the relationships between Tunnicliffe and Bloom and provides examples of questions or comments that could be coded within their learning frameworks. Additionally, the Zoo Cognition Hierarchy may be used to analyze the activities teachers and zoos ask students to complete during a zoo

Table 11.1 Zoo cognition, focused looking, and Bloom's Taxonomy

Zoo cognition hierarchy	Tunnicliffe's levels of focused looking	Bloom's Taxonomy of cognitive gains	Examples of students' interactions at an exhibit
Recognition	"Level 1 is when the focus is on looking at the animals, but in a 'leisure' manner. That means that the conversations are similar in content and orientation to those of non-school visitors"	Knowledge	Hey, look. It's a red panda
Confirmation	"Level 2 is when the school groups do show some focus on a particular theme and look at certain animals, not just wander aimlessly, for example looking at animals from a particular region or looking for 'red' or 'large' animals. However, the quality of the conversations, in terms of the form and content of the dialogue, is still similar to that of non-educationally oriented groups. There is some questioning and answering, but in an information-checking manner"	Comprehension	Yes. It is a red panda. The sign says it lives in China. It says it is vulnerable
Discussion	"Level 3 is when there is some recognisable difference in content and form between the leisure visitors' and school groups' conversations. The topics discussed emphasise and relate to the curriculum and its skills – such as reading, spelling and stories read, as well as topics such as conservation. Groups may use simple keys"	Application	What does that mean? I think it means that it might be endangered. Remember we talked about endangered in class. That means it may disappear
Connection	"Level 4 is when there is identifiable education – new knowledge and progression in conversations occurring in terms of content and of form. Teachers ask learning-questions of the pupils and vice versa, but the way in which the answers are handled and the questions phrased show that this is educational dialogue, not everyday communication. Observations are related to previous knowledge"	Analysis	We need to save the red panda. It is like saving the hawks and eagles. People changed how they lived and saved the birds. They stopped using harmful pesticides. It could be the same
Exploration	"Level 5 is when groups or individuals raise questions, plan how to find the answer, make observations and interpret their data, using previous knowledge as well as their first-hand observations, or discuss what they see and possible explanations"	Synthesis Evaluation	I wonder what people in China are doing to help them? They could save their habitat. How can they save habitat? There are a lot of people in China. Maybe we could write the zoo in China and find out how they save them. I think if they stopped building houses they could save the red panda

visit. If the questions or worksheets do not address the interactions suggested by the Zoo Cognition Hierarchy, higher-order thinking most likely will not occur.

Characteristics of Successful Field Trips

Teachers are the arbitrators and initiators of learning associated with field trips (Tran, 2007; Xanthoudaki, 1998). Therefore, identifying teachers' field trip designs and their decisions to use educational activities are important. Kisiel (2006b) has examined 115 teachers' instructional field trip strategies. Ninety-two percent of the teachers reported using a pre-visit activity. However, 69% reported asking students to complete an activity during the visit. The during-visit activities were identified as (1) structured student engagement strategies, (2) unstructured student engagement strategies, and (3) student supervision. Even though teachers involved students in during-visit activities, the most popular activity was a worksheet.

During a field trip, teachers should provide students with meaningful cognitive and/or affective experiences. The connections teachers make between the field trip and the curriculum influence the cognitive gains, while the holistic experience of the trip shapes the affective gains (Kisiel, 2005; Sheppard, 2000). Developing a learning framework that scaffolds the visit from informational to active learning will provide the teacher and students with a clear, comprehensible purpose. To address the importance of the teachers' decisions concerning a field trip, Davidson, Passmore, and Anderson (2010) have identified and defined four characteristics or implications of a successful field trip design. The characteristics as defined by Davidson et al. are (1) planning, (2) visiting the facility, (3) making the field trip fun, and (4) combining student and teacher-led learning. Good field trip planning can avoid disasters and lead to a successful event (Nabors, Edwards, & Murray, 2009). Moreover, to significantly impact student learning, teachers should incorporate pre-visit, during-visit, and post-visit classroom teaching into the field trip (Hooper-Greenhill, 2000; Kisiel, 2003a; Sheppard, 2000). Davidson et al. (2010) state that maximum classroom input equals maximum field trip gains.

> If they [teachers] want their students to have maximum gains in learning, especially beyond surface learning of facts, teachers need to give students opportunities to build trip learning experiences into classroom activities and ideas, and follow through with these after the trip. If teachers merely rely on a zoo educator to lecture students, or give them a worksheet, student learning will most likely be shallow and fleeting. (p. 138)

Patrick (2008) reviewed the literature concerning field trips and identified three overarching educational concepts of field trip design. Within each educational concept, there are descriptors teachers should consider when developing a successful zoo field trip. The overarching concepts and their descriptors are:

(1) *Cognitive* (A) Pre-visit Activities, (B) During-visit Activities, (C) Post-visit Activities;
(2) *Procedural* (A) Facility Staff, (B) Advanced Organizers;
(3) *Social* (A) Student Groups, (B) Control of Visit and Learning.

Cognitive: Pre-visit Activities

Field trips produce beneficial outcomes when the trips have a purpose and when the students are prepared for the visit. When teachers use focused pre-visit preparation and instruction, there is a positive effect on student learning and attitudes (Falk & Dierking, 2000; Gennaro, 1981), and learning is enhanced (Gennaro, Stoneberg, & Track, 1982; Gross & Pizzini, 1979; Price & Hein, 1991). Students' experiences in informal science learning environments should be teacher focused. Unfortunately, teachers seldom are prepared for field trips, students' excitement, and delineating student learning (Kisiel, 2003b). Teachers do not establish clear, specific objectives for visits to places of informal science learning. Moreover, there is usually little preparation for the visit and little monitoring of learning during the visit (Kisiel), leaving students' with questions concerning how the field trip relates to the classroom. Indeed, children's descriptions of what they learn during a museum visit are based on their prior knowledge, interests, and sociocultural backgrounds. Students do not necessarily link their classroom-based experiences, the curriculum that teachers teach, the pre-visit classroom activities, and the educational objectives with their museum visit (Anderson, Piscitelli, Weier, Everett, & Taylor, 2002; Storksdieck, 2001). Therefore, teachers need to be aware of pre-visit classroom interactions and students' prior knowledge, foci, interactions, and reactions during a field trip, so they more effectively may design field trip experiences.

Students "who visited arts-based museums and engaged in classroom experiences where specific and directly linked content, process, and vocabulary were introduced prior to a museum visit" (Anderson et al., 2002, p. 227) were more engaged in and benefited from the field experience. Prior to the field trip, it is the teacher's responsibility to introduce the purpose and agenda of the field trip to their students. Additionally, it is important for teachers to identify students' possible misunderstandings in relation to the concepts that emerge during the informal science visit (Guisasola, Solbes, Barragues, Morentin, & Moreno, 2009) and to be mindful as to how the novelty of the visit can interfere in learning new information. Therefore, teacher planning prior to the field trip does make a difference in students' post-visit understandings (Guisasola et al., 2009) and increases their learning during the trip (Orion, 1993).

Prior to the zoo visit, teachers may ask students to draw the zoo, talk about the zoo, and write about what they hope to see and their anxieties. For example, students may be concerned that they will have to touch a spider. Allow students to write questions about the visit and place them in a container. This will provide students with anonymity and will allow them to ask better questions.

Cognitive: During-Visit Activities

Three of Tunnicliffe's (1999) zoo visit stages take place at the zoo: concentrated or focused looking, leisure looking, and completion, indicating that the structure of the zoo visit itself is of utmost importance. Traditionally, to incorporate science, teachers have assigned task-oriented activities (Kisiel, 2010), extended an activity already undertaken in class, and/or relied on the interactions of students and chaperones (Griffin & Symington, 1997). However, field trips need to integrate problem-solving skills (McLoughlin, 2004), tie into the curriculum, focus on the standards, take into consideration the children's needs (Nabors et al., 2009), and be supported by appropriate pre- and post-visit activities. During-visit activities should be defined by explicit learning goals and reinforced by the institution's personnel (Anderson et al., 2000; Bhatia & Makela, 2010; Davidson et al., 2010; Henriksen & Jorde, 2001). Rickinson et al. (2004) determined that if field trip activities are "properly conceived, adequately planned, well taught and effectively followed up," they can offer "learners opportunities to develop their knowledge and skills in ways that add value to their everyday experiences in the classroom" (p. 1).

Many museum visits are conducted in a way that does not encourage learning. When students are not introduced to scientific jargon prior to the visit, asked questions that promote higher-order thinking skills, and involved in hands-on activities (Pace & Tesi, 2004), the during-visit activities will fall short of the teacher's learning goals (Tal & Morag, 2007). A successful field trip design creates numerous meaningful learning opportunities and encourages students to continue to ask questions after the field trip.

Teachers are comfortable with controlled learning in the classroom; therefore, teachers prefer to control students' learning during a zoo field trip. Task-oriented field trip activities are an attempt to control the students. This may be due to a lack of knowledge about effective field trip design, an inability to teach in an informal learning setting, and a feeling of fear or intimidation (Griffin & Symington, 1997). When designing a field trip, teachers utilize what they know. Teachers structure the during-visit activities around worksheets (Kisiel, 2006b, 2007, 2010). The worksheets may be scavenger hunts, crossword puzzles, or open-ended questionnaires. Using worksheets to objectify the field trip experience to the point of students' exhaustion is known as "death by worksheet." The overuse of worksheets may come at the expense of true learning.

> Worksheets interfere with the social interaction and maybe learning... while filling in worksheets, group members spoke less to one another, looked at the exhibits less, usually gave up on the worksheets during their tour and didn't spend any more time in the aquarium than groups without worksheets. (Parsons & Muhs, 1994, p. 60)

Moreover, when students are forced to use worksheets during a visit, the worksheet, not the exhibit, becomes the focus of the visit (Griffin & Symington, 1997). Price and Hein (1991) state that "... worksheets too often actually impede student learning by inhibiting true observation, preventing students from formulating their own questions, and causing students to focus on the narrowly described task to the exclusion of broader questions" (p. 515). Students agree with Price and Hein. When students were interviewed

> ... following a trip to a museum there was discussion about the use of worksheets and two students' views provide insights into the reasons that they did not like them: ... *because*

you have to go around looking for the information, you haven't got time to study the things we want to see. [We didn't learn anything because] most of it went in through the eyes and out through the pen on to the paper. (Griffin, 1998 as cited in Griffin, 2004, p. 63)

Kisiel (2003b) analyzed teacher-prepared, field trip worksheets. The worksheets lacked a connection to the classroom curriculum and were not relevant to classroom learning. However, Kisiel discovered two worksheet styles: concept agenda worksheets and survey agenda worksheets. Teachers who utilized the concept agenda worksheets wanted their students to focus on a specific learning goal and focused on a few exhibits. Conversely, teachers who employed survey agenda worksheets were concerned with students viewing as much as possible, reading identification labels, and gathering information. When given a choice between a concept agenda worksheet and a survey agenda worksheet, teachers tended to choose the survey agenda worksheet (Kisiel, 2006c). Teachers chose the survey agenda worksheet because they thought that it would keep students occupied for longer periods of time. However, when survey agenda worksheets are utilized, "the development of a deeper understanding of a particular concept is lost. By limiting students' choices and ignoring students' interests and connections to prior knowledge, survey agendas, as suggested by Worksheet A [survey-oriented], miss valuable opportunities for student learning" (Kisiel, 2007, p. 39). Teachers were under the impression that a worksheet kept students on-task. Additionally, the teachers thought that if students were not provided with worksheets, learning might not occur. Surprisingly, even though students do not like worksheets, they agreed that students would not learn anything if they did not use worksheets and that just looking around did not count as learning (Griffin, 1994). These studies indicate that teachers are more concerned with student control than the relevance of the information being learned. Goal-oriented survey agenda worksheets that rely on student control rather than learning contribute to the epidemic of "death by worksheet."

However, "death by worksheet" is not inevitable. Field trip worksheets may be structured to enhance learning. Riddle (1980) and McManus (1985) have observed students using worksheets during a visit to London's Natural History Museum. They have found that worksheets lead to students interacting with a larger number of exhibits than they would normally. Worksheets appear to be most effective in promoting basic knowledge acquisition during a field trip (Kromba & Harms, 2008). The use of worksheets with younger students may aid them in locating objects/subjects and lead them to personalizing the experience. For older students, worksheets may enhance their social cohesion, if the worksheets are completed as pairs or in groups, and assist in understanding interactive exhibits (McManus). Kromba and Harms have found that worksheets do increase student knowledge and suggest six qualities of an effective worksheet: (1) frequency and extent, (2) free time, (3) choice, (4) guidance, (5) social interaction and cooperation, and (6) focusing attention. Mortensen and Smart (2007) have found similar results. They have determined that a free-choice field trip worksheet developed around learning goals is effective. Students who use a free-choice field trip worksheet have more varied curriculum-related conversations than groups who do not use a free-choice field trip worksheet. Free-choice worksheets may accomplish the teachers' curriculum goals, zoo's free-choice learning goals, and the students' personal learning goals. Even though worksheets used correctly may promote the acquisition of basic knowledge, they should be used sparingly.

Inquiry-based learning and hands-on activities have gained popularity in classroom education and encourage children to explore their environments.

> Inquiry is an approach to learning whereby students find and use a variety of sources of information and ideas to increase their understanding of a problem, topic, or issue. . . . It espouses investigation, exploration, search, quest, research, pursuit, and study. Inquiry does not stand alone; it engages, interests, and challenges students to connect their world with the curriculum. (Kuhlthau, Maniotes, & Caspari, 2007, p. 2)

When students are more engaged in hands-on activities and free-choice learning during a field trip, they retain more knowledge about the field trip and recall a more positive field trip experience (Pace & Tesi, 2004; Tofield, Coll, Vyle, & Bolstad, 2003). Inquiry-based learning (including hands-on activities and free-choice learning) and museum learning share explorative and connective qualities, making them a valuable and effective partner. Due to their interconnectedness, inquiry-based learning in the classroom can be directly supplemented by inquiry-based zoo visits (Rapp, 2005). Inquiry-based learning occurs during a zoo visit because

> . . . the visitor must actually collect and classify data, compare classifications, synthesize an answer to the question based upon that data, and ultimately take a stand on the question, which is based upon what he truly believes and not upon what someone else has told him. In this type of learning situation, the visitor is developing his own rational powers, and he is simultaneously developing an understanding of the content involved because he is evolving the content as he works. The museum visitor is also learning how to find out the information that he does not have and increasing his confidence in his ability to inquire; in short, the museum visitor is learning how to learn—the ultimate goal of museum education. (Booth, Krockover, & Woods, 1982, p. 17)

Therefore, a goal of field trip design is to involve students in various inquiry-based projects during an in-a-zoo visit (Henson, 2008). The inquiry-based projects allow the students to construct knowledge, use higher-order thinking skills, and make connections to their classroom curriculum.

Cognitive: Post-visit Activities

A successful field trip design includes classroom activities that are completed after the visit and organize, build on, and connect the pre-visit and during-visit activities. For teachers to make the most of a field trip they ". . . must include some sort of follow-up activity. The more connections a teacher can make, the greater the opportunity for supporting student understanding" (Kisiel, 2006c, p. 7). Moreover, the post-visit activities provide the students with an understanding of how the field trip relates to their learning in the informal environment. The post-visit activities are an important aspect of tying together all components of the field trip. When students are engaged in pre-visit, during-visit, and post-visit activities, they think creatively, analytically, and critically. This does not include a fill-in-the-blank worksheet, but does challenge students to interpret new information.

Anderson et al. (2000) have studied the impact of post-visit field trip classroom activities on students' learning and knowledge construction. Throughout the field trip, the students worked on a series of activities such as concept maps and open-ended experiments. Anderson et al. have found that the process of integrating pre-visit, during-visit, and especially post-visit activities provides students with an opportunity to construct and reconstruct their personal knowledge of science concepts and principles. This study correlates with Tunnicliffe's (1999) fifth stage of a zoo visit, consolidation. The consolidation and extension stage identifies the importance of using the zoo visit experience to extend students' learning and analytical abilities.

Even though meaningful post-visit activities that are connected to the curriculum do make a difference in learning (Falk & Dierking, 2000; Kisiel, 2006a; McLoughlin, 2004), they are incorporated into the field trip design less often than pre-visit and during-visit activities (Kisiel, 2010). When organizing a zoo field trip, teachers must realize that the educational possibilities do not end when the students leave the zoo. Instead, prior knowledge and concepts learned before and during the zoo field trip can be refined and developed in meaningful ways by utilizing post-visit activities. Such activities are critical to anchoring the field trip to classroom learning and the curriculum. The lack of suggestions for post-visit activities confirms that this is the weakest link in theory and practice.

To expand the zoo visit upon returning to the classroom, students should collect data during the zoo visit. Once they return to the classroom, students may use the data in numerous ways. Students may take pictures during the visit and design class picture books that describe the visit (Davison, 2009; Scott & Matthews, 2011). Cameras offer students an opportunity to produce visuals of the zoo visit and provide students a new language they may use to explain the visit (Clarke and Moss, 2001). Children like to take photographs of the animals that interest them and the animals' anatomy and behavior, because the animals generate emotional connections. Students can design data posters that display information about their observations at the zoo (Gilbert, Breitbarth, Brungardt, Dorr, & Balgopal, 2010). After the visit, students may design informative multimedia presentations that can be shown to the class or other classes (Vavoula, Sharples, Rudman, Meek, & Lonsdale, 2009; Wishart & Triggs, 2010). Zoos should be encouraged to provide websites where students can display and discuss the data or observations they made during the visit. It is obvious that using data collected during field trips is one of the most important aspects of student learning.

Suggested Activities

The following is a list of topics for each subject area. Some of the topics are cross-curricular in nature and are meant to provide a starting point for zoo and classroom educators.

Language Arts/Literacy: Creative and report writing for grammar, spelling, and punctuation; reading labels and information sheets; library research skills at school and at the zoo; fiction reading; writing précis and poetry; reading books about animals and determining the facts, for example, Charlotte's Web (spiders do not talk or spell); and keeping journals during the visit

Mathematics: Number bonds; recognition of simple geometric shapes; basic measurement; estimating and measuring actual weight, length, and volume; use of measurement in real situations, for example, animal diet, visitor numbers, compass points, and orientation; counting the number of legs on an animal; and addition, subtraction, multiplication, and division

Science: Science as an inquiry process, science knowledge and literacy, food chain, plants roles in different habitats, zoology, taxonomy, behavior, using and constructing taxonomic keys, determining how animals' body parts work, identifying salient features, adaptations to the environment and food, conservation biology, physical science, observing animals and designing a study of their behavior, study light by making periscopes that mimic giraffes, swim bladders using floating and sinking objects, center of gravity like birds and monkeys, earflaps and sound, spreading the load of weight like camels' and elephants' feet, and heat insulation using fat, fur, and feathers

Social Studies (History): Animals in history; animals in heraldry, for example, American Eagle; history of exotic animals and zoos; domestication; animals as food sources; animals and the colonization of the western USA; Native Americans and animals; animals in warfare; and uses of animals in different civilizations

Social Studies (Geography): Biomes; major features of the planet such as land masses, oceans, longitude, and latitude; interpretation of maps and symbol; map reading; orientation using a zoo map; constructing maps; reading maps; longitude and latitude; using GPS to map the zoo and favorite exhibits; weather watch; measuring weather in the zoo; and matching animals' adaptations to endemic habitats

Physical Education: Movement replicating those of animals such as running, climbing, and swimming; different ways of moving such as plantigrade and digitigrade; and comparing physical abilities between the students and the animals—How far can you run in 1 min compared with the cheetah?

Design Technology: Identifying and meeting the animals' needs; visitor questionnaires; and designing new exhibits, food dispensers, labels, and visitor guides

Information Technology: Word processing and data collection, handling, and analyzing

Music: Composing music appropriate for different animals, using instruments to replicate animal sounds, and listening to music written about animals, for example, Peter and the Wolf

Art: Drawing and painting from firsthand observations of animals; learning names of colors found on animals; drawing animal body covering colors and patterns; making scientific observational drawings of animals; making models of animals using modeling clay, clay, or other materials such as papier-mache; and taking photographs and displaying them during an art show

Procedural: Facility Staff

Contacting the zoo's staff and visiting the facility prior to the visit are important aspects of the field trip design. By visiting the facility prior to the field trip, the teacher makes the students' informal experience more effective (Melber, 2000). Moreover, classroom teachers need to work closely with the facility's staff.

> ...they [zoo educators] need to work closely with classroom teachers to ensure there are clear, explicit learning goals, that the zoo educator knows how the trip fits in with pre- and posttrip classroom activities, and what the students want and expect from the trip. (Davidson et al., 2010, p. 138)

Problems occur when teachers do not feel comfortable taking students on field trips or they are not sure what to do during the field trip. This is especially true for new teachers (Kisiel, 2010). This lack of knowledge or feelings of insecurity may be satiated by interacting with the informal educators. Indeed, by interacting with staff and chaperones, teachers can make a difference in learning (Parsons, 2010; McLoughlin, 2004). However, teachers should not rely directly on docent-led tours, because docent-led tours do not offer significant inquiry-based learning opportunities (Cox-Petersen et al., 2003). Moreover, student learning is affected directly by the teacher's learning agenda and views on learning, but does not seem to be affected significantly by zoo educator agendas (Davidson et al., 2010). Docents and zoo educators do not have significant effects on student learning (Davidson et al.). This may be due to the fact that students spend more time with their classroom teachers and therefore understand the teacher's learning agenda. This suggests that if informal zoo educators are going to affect student learning, then the zoo educator's agenda must flow through the classroom teacher. Interactions between the classroom teachers and the informal educators may positively influence informal educators and docents. When both educators understand each other and their educational goals, learning during a zoo field trip will be more likely to occur. Therefore, communication between informal educators and classroom teachers is important.

However, the communication between informal educators and classroom teachers appears to be poor (Noel, 2007). Additionally, a disconnect exists between what informal educators see as their role in field trips and their actual roles. Informal educators often view their roles as motivational (Noel). Nevertheless, informal educators need to perceive themselves as educational collaborators who work with teachers in schools to provide the best educational experiences to children during a field trip. Informal educators and classroom teachers can work together to insure that children are prepared for a field trip, have field trip related materials, and engage in a quality educational experience (Noel & Colopy, 2006).

Research suggests that teachers need to speak with the facility's informal educators and invite the informal educators into the classroom (Anderson & Lucas, 1997; Melber, 2000). Conversely, informal educators need to develop a relationship with classroom teachers and use their staff and exhibits to provide successful informal education experiences (Myers, Stanoss, Jenke, & Stowell, 2009). The relationships that develop between classroom teachers and informal educators are a key to building successful field trips (Xanthoudaki, 1998). A successful partnership between a school and the field trip destination requires significant communication and members of both organizations taking the time to learn about each other (Kisiel, 2010). Once a strong relationship is cultivated, it can provide immense benefits for teachers and students. By working together, classroom teachers and informal educators can build a culture of inquiry (Myers et al., 2009).

Even when schools arrange field trips to the zoo, the collaboration between the school and the zoo frequently does not extend beyond the visit. However, there are examples of schools and zoos working together in partnerships to create unique and rich learning environments for students. The Lincoln Children's Zoo in Lincoln, Nebraska, has been involved in two classroom-based learning projects: Zoo to You and Bumble Boosters. The Zoo to You program loans zoo animals to classrooms during the winter months when the zoo is closed. Participating classrooms have four different animals for six weeks each. During the six weeks, the students file reports with the zoo regarding the animals' health and wellness. Additionally, students use inquiry learning to develop their knowledge of the animals and to keep detailed journals of their observations. Students develop inquiry-based experiments and post them on the zoo's website. Once the zookeeper approves the experiment, students implement the research design. For example, students test the animal's food preferences by providing the animals with different food choices. The students' learning is assessed through students' journal entries, Venn diagrams, and self-evaluations (Wickless et al., 2003). The goal of Bumble Boosters is to involve students in authentic scientific research and improve science-inquiry skills. The project had two goals: (1) to determine the distribution and floral preferences of Nebraska bumblebee species and (2) to design artificial hives that attract bumblebee queens. Students collected over 3,200 bumblebees, built 65 artificial hives, and collected one species of bumblebee that had not been previously seen in Nebraska (Golick, Schlesselman, Ellis, & Brooks, 2003). The teachers who have participated in Zoo to You and Bumble Boosters have seen improvements in the students' ability to ask inquiry-based questions and apply the scientific process (Wickless et al.). Teachers believe that the zoo-related projects provide richer opportunities for students to use the scientific process than classroom experiments, promote the generalization of inquiry questioning to other science topics, and build students' confidence in their abilities as researchers (Golick et al., 2003). Moreover, students have shown an increase in the use of scientific facts and science vocabulary in their expository writing (Trainin, Wilson, Wickless, & Brooks, 2005).

Procedural: Advanced Organizers

Zoo visitors arrive with differing agendas, reasons for the visit, and backgrounds. Even students' agendas differ from their teachers' ideas concerning the field trip experience (Anderson et al., 2002; Storksdieck, 2001). Teachers usually overlook these competing agendas, but they are an important part of planning a successful field trip (Anderson et al., 2008). Moreover, the field trip site itself can inhibit learning due to the novelty of the setting. The visit to an unfamiliar location and the process of integrating new stimuli easily can overwhelm the students (Falk et al., 1978). When students are more familiar with the field trip setting, they are more likely to learn, stay on task, and benefit from teacher-planned activities (Tunnicliffe, 1999). During a field trip, students who are not familiar with the field trip setting are often off-task and engage in behaviors that inhibit the teacher-planned learning activity. Additionally, Falk and Balling (1982) have revealed that the age of the students may affect the novelty level at which optimal learning will take place. They show that younger students may be influenced more easily by the novelty of the field trip and consequently will have less significant knowledge gains.

Additionally, visitors with minimal to moderate interest showed significant gains in conservation interest and concern. The researchers explain the results by saying that "Most experts do not find museum-like settings ideal for dramatically furthering their knowledge. Museums, zoos, and aquariums are designed primarily to attract, engage, and stimulate visitors with limited knowledge and at least moderate levels of interest" (Falk & Adelman, 2003, p. 172). The implication of these results for a teacher organizing a zoo field trip is that his or her primary goal in preparing students for the visit should be to encourage their interest in the subject as much as possible to achieve optimal learning. Prior knowledge is also important, but less so than interest level. In essence, knowledge can be gained at the zoo, but interest cannot, and teachers should focus on creating interest about the zoo field trip and animals when preparing their students.

The preparation of students for a field trip is one of the most important aspects of organizing a trip to the zoo that will result in learning. Some degree of novelty is conducive for learning; therefore, teachers must take into account the age and background of their students (Falk & Balling, 1982). These results indicate teachers should prepare students with as much knowledge about the field trip setting as possible to reduce the novel field trip phenomenon and increase learning from structured activities. Teachers should capitalize on the innate curiosity of students in novel environments and use it to help them learn (Falk et al., 1978). However, not only should children be prepared for the novelty of the physical setting but also for the content and information they will encounter. In order to prevent students misinterpreting the reasons for the visit, teachers may provide graphic organizers (McLoughlin, 2004) and field trip objectives (Skop, 2008).

The advanced field trip organizer is a packet of information the teacher provides the students and chaperones prior to the zoo visit. Zoo field trip organizers should include (at least): (1) the visit agenda with times for arrival, lunch, and departure;

(2) a list of the common and scientific names (This is for the teacher; students may not be interested in the scientific name) of the plants and animals that students might observe; (3) the science concepts to which students should pay attention; 4) a map of the zoo; and 5) a description of the zoo. The advanced organizer should include a discussion of the teacher's expectations. Discussing the advanced organizer provides an opportunity to discuss the route, students' behavior, lunch, and shop visits. However, field trip organizers are beneficial only if the students and chaperones review the organizer before the visit. Field trip organizers should not be handed to students and chaperones on the bus the morning of the visit or upon entrance to the zoo. Sharing advanced field trip organizers with students and teachers may involve showing slides or a video, providing the groups with a timetable, and discussing opportunities for visits to the gift shop, when and where lunch is to be taken, and "housekeeping" issues. During the advanced field trip organizer session, provide students with zoo maps and pose various problem-solving questions. Ask students to use orientation skills by example posing questions such as, "What route would you take to reach the Aviary from the chimpanzees?" This is also a time to ask students to think about observing animals instead of viewing animals. For example, provide students with pictures of animals and ask them to describe the animal and the behavior(s) they are exhibiting in the photograph. If students learn to observe the animals before the zoo visit, they will be more perceptive of the animals during the visit.

Social: Student Groups

Field trips should be fun for students (Hamilton-Ekeke, 2007; Nabors et al., 2009). Griffin (2004) has found that when students are given more choices in the field trip design, they have fun and do things they like but still view the trip as educational. However, field trips are weighted with responsibility and obligation to learn, brought on by the demands of worksheets and specified activities. There is no room in the field trip for fun. Objectives must be met and the field trip must accomplish prespecified goals. Moreover,

> Curriculum and teaching sometimes serve to kill student interest, debase the child, stymie natural curiosity. . . . Many of us, especially in schools, objectify things that are as pure, as natural, and as expressive as we would expect a zoo trip to be. We needlessly turn such wonderful experiences — not meant as structured, organized learning activities — into labor. (Poetter, 2006, p. 323)

For many students, the most important part of a field trip is the social aspect. The students' social interactions enhance learning activities (Davidson et al., 2010). "This could be brought about by allowing students to be in groups with their friends and could be focused by having students discuss what they saw, learned, and enjoyed with each other . . ." (Davidson et al., p. 138). Additionally, if teachers allow students to define the groups, the teacher should still pay attention to gender, class, classroom history, and prior knowledge (Bätz, Wittler, &Wilde, 2010; Neff, 1977; Skop, 2008; Thomson, Buchanan, & Schwab, 2006). Even though teachers are concerned about maintaining discipline and control during the field trip (Kisiel, 2010) and believe that students are not capable of making good grouping decisions, allowing students to choose their own working groups gives the students a feeling of control, may lessen the teacher's workload, and may increase students' discussions.

Social: Control of Visit and Learning

The best field trips involve elements of both instructor-led explanations and student-centered exploration/discovery. Teachers should allow students to make suggestions about the field trip design (McLoughlin, 2004). Allowing students control and choice concerning the field trip encourages engagement and motivation (Kromba & Harms, 2008).

> Perhaps alternative theoretical perspectives, such as critical theory, could expand on how such engagement could be obtained. Critical pedagogy urges us to be aware of how power relationships in the classroom affect what students learn; where the teacher, a dominant authority figure, imposes values and structure ... when this relationship is made explicit, teachers can use their position to help students participate in their education—give them a voice .... (Davidson et al., 2010, p. 139)

Providing students with choices concerning the exhibits they visit and knowledge they acquire allows students to feel empowered in the relation to their own learning. When students are asked how they would design a field trip, they say:

> I'd give the children the choice of what they want to do and join everyone together and find out what they want to learn. Make it for the whole day .... And let us go where we want and find out what we want to find out. We'd organize it for longer. We'd be able to go around and do it ourselves, not as many questions, be able to see other things. (Griffin, 1998 as cited in Griffin, 2004, p. 60)

Because zoos are places of informal learning, they inherently foster free-learning choices. These choices are available to non-field trip visitors, but are not traditionally an element of a field trip design. Non-field trip visitors "construct personal meaning, have genuine choices, encounter challenging tasks, take control over their own learning, collaborate with others, and feel positive about their efforts" (Paris, Yambor, & Packard, 1998, p. 271). Even though the successful aspect of choice during a non-classroom-related trip has been determined, choice does not seem to be a priority in field trip designs. By allowing students a choice in designing the field trip agenda and choosing educational topics, the field trip will become a personal expansion of learning interactions.

Teacher Training and Chaperone Preparation

When teachers and chaperones are excited about learning, active with students during the visit, display personal interest in the visit, and are prepared for learning, students show more interest in the visit and have positive attitudes toward the zoo visit (Jarvis & Pell, 2005). Therefore, teachers and chaperones are an important aspect of the zoo visit experience. Teacher training and chaperone preparation differ from field trip design. Teachers need to be trained in the techniques of designing a successful field trip, and chaperones need to be prepared to interact with students during the field trip. For a field trip to be a successful educational out-of-classroom experience, teachers and chaperones must understand their roles. Chaperones should be provided with advanced organizers that include zoo maps, an agenda, any activities in which students will participate, guiding questions (especially for young children), emergency information, and rules of behavior (for chaperones and the students). Preservice teachers who listened to student groups' conversations during a zoo field trip observed chaperones smoking even though the zoo was a nonsmoking facility (Patrick, 2008).

Before teachers are suited better to prepare chaperones, they must themselves understand field trip design. Though the field trip provides a learning experience different from that of the classroom, many teachers design field trips around worksheets (Kisiel, 2005) and do not focus on the pedagogical aspects of learning during a field trip. Teachers are disconnected from the field trip setting (Kisiel, 2010), their pedagogical understanding of field trip design is not developed fully, and their conceptions regarding informal learning are not developed. Classroom teachers and preservice teachers need educational opportunities to develop their field trip objectives and understandings of field trip design. Therefore, educators in formal and informal settings need to "consider different means of teacher support that may help reduce apprehensions and shape attitudes regarding what a successful excursion might look like" (Kisiel, 2007, p. 41) and help teachers "to become more aware of the characteristics of these nonclassroom settings that facilitate learning, such as visitor choice and control" (Kisiel, p. 41).

Because preservice teachers have had little or no training on how to incorporate informal learning environments into the curriculum, there is a need to educate them about conducting informal learning experiences (Melber, 2000). Moreover, involving preservice teachers in field trips could motivate them to take students on field trips (Munakata, 2005). Johnson and Chandler (2009) state that

> ...empowering pre-service teachers with the knowledge of how to effectively execute a field trip will only help ease the possible anxiety these teachers may face when planning a field trip themselves. Furthermore, pre-service teachers possess the unique ability to view an informal learning environment from both the student and the teacher's perspective. Having pre-service...teachers...experience such events forces them to think about what constitutes a productive field trip before they ever enter a classroom as well as consider components of the curriculum that would benefit from an environment-influenced education. (p. 8)

University teacher educators should include field trip design in their preservice teachers' methods courses. Teacher educators need to "help teachers reflect more carefully on their pedagogy, regardless of the location of the lesson" (Kisiel, 2007, p. 41). Providing preservice teachers with opportunities to develop science lessons, which incorporate informal science visits (Bulunuz & Jarrett, 2010), will encourage them to take students to places of informal learning (Patrick, Matthews, & Tunnicliffe, 2011).

While teachers should familiarize themselves with the site ahead of time and define the learning experience (Kisiel, 2006a), the role of the zoo educator is to support classroom teachers. Informal education sites should offer continuing education opportunities for classroom teachers. Because classroom teachers affect the positive gains students have toward science (Jarvis & Pell, 2005), zoo educators should utilize the classroom teacher as a conduit to reach the student visitor. Teachers who participate in zoo education programs positively influence their students' knowledge and science attitudes (de White & Jacobson, 1994). This suggests that zoo educators can impact students by addressing teachers as their primary audience through inquiry-based workshops, seminars, and educational materials.

References

Anderson, W. (1973). Is their journey really necessary? *The Times Educational Supplement, 3041*, 41–42.

Anderson, D., & Lucas, K. (1997). The effectiveness of orienting students to the physical features of a science museum prior to visitation. *Research in Science Education, 27*(4), 485–495.

Anderson, D., Lucas, K. B., Ginns, I. S., & Dierking, L. D. (2000). Development of knowledge about electricity and magnetism during a visit to a science museum and related post-visit activities. *Science Education, 84*(4), 658–679.

Anderson, D., Piscitelli, B., & Everett, M. (2008). Competing agendas: Young children's museum field trips. *Curator: The Museum Journal, 51*(3), 253–273.

Anderson, D., Piscitelli, B., Weier, K., Everett, M., & Taylor, C. (2002). Children's museum experiences: Identifying powerful mediators of learning. *Curator, 45*(3), 213–231.

Association of Science Education. (2001). *Be safe! Health and safety in primary school science and technology* (3rd ed.). Hatfield, UK: Association for Science Education.

Ballantyne, R., Packer, J., Hughes, K., & Dierking, L. (2007). Conservation learning in wildlife tourism settings: Lessons from research in zoos and aquariums. *Environmental Education Research, 13*(3), 367–383.

Bätz, K., Wittler, S., & Wilde, M. (2010). Differences between boys and girls in extracurricular learning settings. *International Journal of Environmental & Science Education, 5*(1), 51–64.

Berry, N. W. (1998). A focus on art museum/school collaborations. *Art Education, 51*(2), 8–14.

Bhatia, A., & Makela, C. (2010). *Making the most of museum field trips*. In Creating the inclusive museum. Edinburgh, UK: Hudson House.

Booth, J. H., Krockover, G. H., & Woods, P. R. (1982). *Creative museum methods and educational techniques*. Springfield, IL: Thomas Books.

Bowers, C. A. (1997). *The culture of denial: Why the environmental movement needs a strategy for reforming universities and public schools*. Albany, NY: State University of New York Press.

Braund, M., & Reiss, M. (2006). Towards a more authentic science curriculum: The contribution of out-of-school learning. *International Journal of Science Education, 28*(12), 1373–1388.

Bulunuz, M., & Jarrett, O. (2010). Developing an interest in science: Background experiences of preservice elementary teachers. *International Journal of Environmental & Science Education, 5*(1), 65–84.

Clark, E. (1943). An experimental evaluation of the school excursion. *Journal of Experimental Education, 12*, 10–19.

Cox-Petersen, A. M., Marsh, D. D., Kisiel, J., & Melber, L. M. (2003). Investigation of guided school tours, student learning, and science reform recommendations at a museum of natural history. *Journal of Research in Science Teaching, 40*(2), 200–218.

Curtis, D. (1944). The contributions of the excursion to understanding. *Journal of Educational Research, 38*, 201–211.

Davidson, S. K., Passmore, C., & Anderson, D. (2010). Learning on zoo field trips: The interaction of the agendas and practices of students, teachers, and zoo educators. *Science Education, 94*(1), 122–141.

Davison, S. (2009, January). A picture is worth 1,000 words: Using digital cameras captivates second-grade learners at the zoo. *Science and Children, 46*(5), 36–39.

de White, T. G., & Jacobson, S. K. (1994). Evaluating conservation education programs at a South American zoo. *Journal of Environmental Education, 25*(4), 18–22.

Falk, J. H., & Adelman, L. M. (2003). Investigating the impact of prior knowledge and interest on aquarium visitor learning. *Journal of Research in Science Teaching, 40*(2), 163–176.

Falk, J. H., & Balling, J. D. (1982). The field trip milieu: Learning and behavior as a function of contextual events. *Journal of Educational Research, 76*(1), 22–28.

Falk, J. H., & Dierking, L. (1992). *The museum experience*. Washington, DC: Whalesback Books.

Falk, J. H., & Dierking, L. D. (1997). School field trips: Assessing their long-term impact. *Curator, 40*(3), 211–218.

Falk, J. H., & Dierking, L. D. (2000). *Learning from museums: Visitor experience and the making of meaning*. Walnut Creek, CA: Alta Mira Press.

Falk, J. H., Martin, W. W., & Balling, J. D. (1978). The novel field-trip phenomenon: Adjustment to novel settings interferes with task learning. *Journal of Research in Science Teaching, 15*(2), 127–134.

Falk, J. H., Reinhard, E. M., Vernon, C. L., Bronnenkant, K., Deans, N. L., & Heimlich, J. E. (2007). *Why zoos & aquariums matter: Assessing the impact of a visit*. Silver Spring, MD: Association of Zoos and Aquariums.

Gennaro, E. (1981). The effectiveness of using previsit instructional materials on learning for a museum field trip experience. *Journal of Research in Science Teaching, 18*(3), 275–279.

Gennaro, E., Stoneberg, S. A., & Track, S. (1982). Chance or prepared mind? *Journal of Museum Education, 7*(4), 16–18.

Gilbert, L., Breitbarth, P., Brungardt, M., Dorr, C., & Balgopal, M. (2010, February). The view at the zoo: Using a photographic scavenger hunt as the basis for an interdisciplinary field trip. *Science Scope, 33*(6), 52–55.

Golick, D. A., Schlesselman, D. M., Ellis, M. D., & Brooks, D. W. (2003). Bumble Boosters: Students doing real science. *Journal of Science Education and Technology, 12*(2), 149–152.

Griffin, J. (1994). Learning to learn in informal settings. *Research in Science Education, 24*, 121–128.

Griffin, J. (1998). School-museum integrated learning experiences in science: A learning journey. Unpublished Ph.D. thesis, University of Technology, Sydney, Australia.

Griffin, J. (2004). Research on students and museums: Looking more closely at the students in school groups. *Science Education, 88*, 59–70.

Griffin, J., & Symington, D. (1997). Moving from task-oriented to learning-oriented strategies on school excursions to museums. *Science Education, 81*(6), 763–779.

Gross, M., & Pizzini, E. (1979). The effects of combined advance organizers and field experiences on environmental orientations of elementary school children. *Journal of Research in Science Teaching, 16*(4), 325–331.

Guisasola, J., Solbes, J., Barragues, J., Morentin, M., & Moreno, A. (2009). Students' understanding of the special theory of relativity and design for a guided visit to a science museum. *International Journal of Science Education, 31*(15), 2085–2104.

Hamilton-Ekeke, J. (2007). Relative effectiveness of expository and field trip methods of teaching on students' achievement in ecology. *International Journal of Science Education, 29*(15), 1869–1889.

Harvey, H. W. (1951). An experimental study of the effect of field trips upon the development of scientific attitudes in a ninth grade general science class. *Science Education, 35*(5), 242–248.

Henriksen, E., & Jorde, D. (2001). High school students' understanding of radiation and the environment: Can museums play a role? *Science Education, 85*(2), 189–206.

Henson, K. (2008). It's a zoo out there! Students conduct content-rich scientific inquiry using authentic ethological methods. *The Science Teacher, 75*(2), 44–47.

Hooper-Greenhill, E. (2000). *Museums and the interpretation of visual culture.* New York: Routledge.

Howland, A. (1962). How to conduct a field trip. *National Council for the Social Studies, How to Do It Series, No 12.* Washington, DC: National Education Association.

Jarvis, T., & Pell, A. (2005). Factors influencing elementary school children's attitudes toward science before, during, and after a visit to the UK National Space Center. *Journal of Research in Science Teaching, 42*(1), 53–83.

Johnson, D., & Chandler, F. (2009). Pre-service teachers' fieldtrip to the battleship: Teaching and learning mathematics through an informal learning experience. *Issues in the Undergraduate Mathematics Preparation of School Teachers: The Journal, 2,* 1–9.

Kellert, S. R. (1985). Attitudes towards animals: Age-related development among children. *The Journal of Environmental Education, 16*(3), 29–39.

Kisiel, J. F. (2003a). *Revealing teacher agendas: An examination of teacher motivations and strategies for conducting museum field trips.* Unpublished dissertation, University of Southern California, Los Angeles.

Kisiel, J. F. (2003b). Teachers, museums and worksheets: A closer look at a learning experience. *Journal of Science Teacher Education, 14*(1), 3–21.

Kisiel, J. (2005). Understanding elementary teacher motivations for science fieldtrips. *Science Education, 89*(6), 936–955.

Kisiel, J. (2006a). More than lions and tigers and bears: Creating meaningful field trip lessons. *Science Activities, 43*(2), 7–10.

Kisiel, J. (2006b). An examination of fieldtrip strategies and their implementation within a natural history museum. *Science Education, 90*(3), 434–452.

Kisiel, J. (2006c). Examining teacher choices for science museum worksheets. *Journal of Science Teacher Education, 18,* 29–43.

Kisiel, J. (2007). Examining teacher choices for science museum worksheets. *Journal of Science Teacher Education, 18*(1), 29–43.

Kisiel, J. F. (2010). Exploring a school-aquarium collaboration: An intersection of communities of practice. *Science Education, 94*(1), 95–121.

Krathwohl, D. (2002). A revision of Bloom's Taxonomy: An overview. *Theory into Practice, 41*(4), 212–264.

Krepel, W. J., & Duvall, C. R. (1973). A study of school board policies and administrative procedures for dealing with field trips in school districts in cities with populations over 100,000 in the United States. *Research in Education, 8*(1), 30–41.

Krepel, W. J., & Duvall, C. R. (1981). *Field trips: A guide for planning and conducting educational experiences.* Washington, DC: National Education Association.

Kromba, A., & Harms, U. (2008). Acquiring knowledge about biodiversity in a museum: Are worksheets effective? *Journal of Biological Education, 42*(4), 157–163.

Kuhlthau, C. C., Maniotes, L. K., & Caspari, A. K. (2007). *Guided inquiry: Learning in the 21st century.* Westport, CT: Libraries Unlimited.

Lockheed, M. E. (1985). *Sex & ethnic differences in middle school mathematics, science and computer science: What do we know?* Princeton, NJ: Educational Testing Service.

Mallapur, A., Waran, M., & Sinha, A. (2008). The captive audience: The educative influence of zoos on their visitors in India. *International Zoo Yearbook, 42*, 214–224.

McLaughlin, E., Smith, W. S., & Tunnicliffe, S. D. (1998). Effect on primary level students of in service teacher: Education in an informal science setting. *Journal of Science Teaching, 9*(2), 123–142.

McLoughlin, A. (2004). Engineering active and effective field trips. *The Clearing House, 77*, 160–163.

McManus, P. (1985). Worksheet-induced behaviour in the British Museum (Natural History). *Journal of Biological Education, 19*(3), 237–242.

Melber, L. (March 2000). Tap into informal science learning. *Science Scope, 23*, 28–31.

Mortensen, M. F., & Smart, K. (2007). Free-choice worksheets increase students' exposure to curriculum during museum visits. *Journal of Research in Science Teaching, 44*(9), 1389–1414.

Munakata, M. (2005). Exploring mathematics outside the classroom through the field trip assessment. *PRIMUS, 15*, 117–123.

Myers, O. E., Jr., Saunders, C. D., & Garrett, E. (2004). What do children think animals need? Developmental trends. *Environmental Education Research, 10*(4), 545–562.

Myers, C., Stanoss, R., Jenke, D., & Stowell, S. (2009, September 12–18). *Advanced inquiry: Deepening engagement in science and conservation at zoos & aquariums*. Paper presented at the Association for Zoos and Aquariums.

Nabors, M., Edwards, L., & Murray, R. (2009). Making the case for field trips: What research tells us and what site coordinators have to say. *Education, 129*(4), 661–667.

Neff, T. R. (1977). The use of field trip courses to strengthen undergraduate geology programs. *Journal of Geological Education, 25*, 26–28.

Noel, A. (2007). Elements of a winning field trip. *Kappa Delta Pi Record, 44*(1), 42–44.

Noel, A., & Colopy, M. (2006). Making history field trips meaningful: Teachers' and site educators' perspectives on teaching materials. *Theory and Research in Social Education, 34*(3), 553–568.

Orion, N. (1993). A model for the development and implementation of field trips as an integral part of the science curriculum. *School Science and Mathematics, 93*(6), 325–331.

Orion, N., & Hofstein, A. (1991). The measurement of students' attitudes towards scientific field trips. *Science Education, 75*(5), 513–523.

Pace, S., & Tesi, R. (2004). Adult's perception of field trips taken within grades K-12: Eight case studies in the New York metropolitan area. *Education, 125*(1), 30–40.

Paris, S. G., Yambor, K. M., & Packard, B. (1998). Hands-on biology: A museum-school-university partnership for enhancing students' interest and learning in science. *The Elementary School Journal, 98*(3), 267–289.

Parsons, C. (2010, September 12–16). *A program can change the entire field trip*. Paper presented at the Association of Zoos and Aquariums, Houston, TX.

Parsons, C., & Muhs, K. (1994). Field trips and parent chaperones: A study of self-guided school groups at the Monterey Bay Aquarium. *Visitor Studies: Theory, Research and Practice, 7*(1), 57–61.

Patrick, P. (2008, October). *Going to the zoo, zoo, zoo…* Paper presented at the National Association of Biology Teachers Conference, Memphis, TN.

Patrick, P. (2011, April). *Zoo acuity model: Middle level students' knowledge of zoos*. Paper presented at the National Association for Research in Science Teaching, Orlando, FL.

Patrick, P., & Getz, A. (2008). Becoming a spider scientist. *Science & Children, 46*(3), 5–10.

Patrick, P., Matthews, C., & Tunnicliffe, S. (2011). Using a field trip inventory to determine if listening to elementary school students' conversations, while on a zoo field trip, enhances preservice teachers' abilities to plan zoo field trips. *International Journal of Science Education*. Available at: http://www.tandfonline.com/doi/abs/10.1080/09500693.2011.620035, 1–25, iFirst article.

Patrick, P., & Tunnicliffe, S. D. (2011). What plants and animals do early childhood and primary students' name? Where do they see them? *Journal of Science Education and Technology*. Invited article and Special Issue: Early Childhood and Nursery School Education. Available at: http://www.springerlink.com/content/e27121057mqr8542/

Pedretti, E. G. (2002). T. Kuhn meets T. Rex: Critical conversations and new directions in science centres and science museums. *Studies in Science Education, 37*, 1–42.

Perry, D. L. (1992). Designing exhibits that motivate. *Association of Science-Technology Centers Newsletter*, 20(1), 9–10, 12.

Perry, D. L. (1993). Beyond cognition and affect: The anatomy of a museum visit. In A. Benefield, S. Bitgood, H. Shettel, D. Thompson, & R. Williams (Eds.), *Visitor studies: Theory, research, and practice, Vol. 6, Proceedings of the 1993 visitor studies conference*. Jacksonville, AL: Center for Social Design.

Poetter, T. S. (2006). The zoo trip: Objecting to objectives. *Phi Delta Kappan, 88*(4), 319–323.

Price, S., & Hein, G. E. (1991). More than a field trip: Science programmes for elementary school groups at museums. *International Journal of Science Education, 13*(5), 505–519.

Rapp, W. H. (2005). Inquiry-based environments for the inclusion of students with exceptional learning needs. *Remedial and Special Education, 26*(5), 297–310.

Rebar, B. (2010). *Teachers' sources of knowledge for field trip practices*. Paper presented at The National Association for Research in Science Teaching 2010 Annual Meeting, Philadelphia.

Rebar, B. M., & Enochs, L. G. (2010). Integrating environmental education field trip pedagogy into science teacher preparation. In A. Bodzin, S. Weaver, & B. Klein (Eds.), *The inclusion of environmental education in science teacher education* (pp. 111–126). London: Springer.

Rickinson, M., Dillon, J., Teamey, K., Morris, M., Choi, M. Y., Sanders, D., & Benefield, P. (2004). *A review of research on outdoor learning*. Preston Montford, Shropshire, UK: Field Studies Council.

Riddle, W. (1980). *A study of adolescent visitors to the human biology exhibition*. Unpublished Master's thesis, Chelsea College, London.

Rosenfeld, S. (1980). *Informal learning in zoos: Naturalistic studies on family groups*. Unpublished doctoral dissertation, University of California, Berkeley.

Ruth, F. (1962). Field trips—Why and how? *The American Biology Teacher, 24*, 410–412.

Scott, C. M., & Matthews, C. E. (2011). The "science" behind a successful field trip to the zoo. *Science Activities: Classroom Projects and Curriculum Ideas, 48*(1), 29–38.

Sheppard, B. (Ed.). (2000). *Building museum & school partnerships*. Harrisburg, PA: Pennsylvania Federation of Museums and Historical Organizations.

Skop, E. (2008). Creating field trip-based learning communities. *Journal of Geography, 107*(6), 230–235.

Storksdieck, M. (2001). Differences in teachers' and students' museum field-trip experiences. *Visitor Studies Today, 4*(1), 8–12.

Tal, T., & Morag, O. (2007). School visits to natural history museums: Teaching or Enriching? *Journal of Research in Science Teaching, 44*(5), 747–769.

Talboys, G. K. (1996). *Using museums as an educational resource: An introductory handbook for students and teachers*. Brookfield, VT: Ashgate Publishing Company.

Thomson, J., Buchanan, J., & Schwab, S. (2006). An integrative summer field course in geology and biology for K-12 instructors and college and continuing education students at Eastern Washington University and beyond. *Journal of Geoscience Education, 54*(5), 588–595.

Tofield, S., Coll, R. K., Vyle, B., & Bolstad, R. (2003). Zoos as a source of free choice learning. *Research in Science & Technological Education, 21*(1), 67–99.

Trainin, G., Wilson, K., Wickless, M., & Brooks, D. (2005). Extraordinary animals and expository writing: Zoo in the classroom. *Journal of Science Education and Technology, 14*(3), 299–304.

Tran, L. U. (2007). Teaching science in museums: The pedagogy and goals of museum educators. *Science Education, 91*(2), 278–297.

Tunnicliffe, S. D. (1992). Zoo education. *International Zoo News Journal, 39*(3), 15–22.

Tunnicliffe, S. D. (1993). *We're all going to the zoo tomorrow, zoo tomorrow: Children's conversations at animal exhibits at London and St. Louis Zoos*. Paper given at Visitor Studies Association Conference, Albuquerque, New Mexico.

Tunnicliffe, S. (1994). Attitudes of primary school children to animals in a zoo. In J. Nicholson & A. Podberscek (Eds.), *Issues in research in companion animal studies*. Study no. 2. Glasgow, UK: Society for Companion Animal Studies.

Tunnicliffe, S. D. (1999). Stages of a zoo visit. *International Zoo News, 46*(6), 343–346.

Tunnicliffe, S., Lucas, A., & Osborne, J. (1997). School visits to zoos and museums: A missed educational opportunity? *International Journal of Science Education, 19*(9), 1039–1056.

Vavoula, G., Sharples, M., Rudman, P., Meek, J., & Lonsdale, P. (2009). Myartspace: Design and evaluation of support for learning with multimedia phones between classrooms and museums. *Computers & Education, 53*, 286–299.

Wagoner, B., & Jensen, E. (2010). Science learning at the zoo: Evaluating children's developing understanding of animals and their habitats. *Psychology and Society, 3*(1), 65–76.

Weiss, I. (1997). The status of science and mathematics teaching in the United States. *National Institute for Science Education, 1*(3), 1–8.

Wickless, M., Brooks, D. W., Abuloum, A., Mancuso, B., Heng-Moss, T. M., & Mayo, L. (2003). Our zoo to you: A zoo's "animal loan" program lets students and teachers experience the benefits of observing animals up close without the worry. *Science and Children, 41*(1), 36–39.

Wishart, J., & Triggs, P. (2010). MuseumScouts: Exploring how schools, museums and interactive technologies can work together to support learning. *Computers & Education, 54*(3), 669–678.

Wolins, I. S., Jensen, N., & Ulzheimer, R. (1992). Children's memories of museum field trips: A qualitative study. *Journal of Museum Education, 17*(2), 17–27.

Xanthoudaki, M. (1998). Is it always worth the trip: The contribution of museum and gallery educational programmes to classroom art education. *Cambridge Journal of Education, 28*(2), 181–195.

Chapter 12
Conclusions

The purpose of this book is to provide classroom and zoo educators with a theoretical framework upon which they design field trips. We have identified and defined two voices: Zoo Voice and Visitor Voice. In the case of this book, the Visitor Voice is considered to be people who visit zoos such as teachers, parents, chaperones, families, leisure groups, and students. As defined in the previous chapters, the Zoo Voice is provided by the zoo. The overlap of these voices potentially may be the space where the Zoo Voice and Visitor Voice fuse in the construction of mutual understanding. This fusion would mean each voice would be aware of and heeding intellectually the other voice. However, this integrated space of overlap is Zoo Noise, because most zoo visitors are not comprehending the Zoo Voice and many zoos are ignoring the Visitor Voice. Why is it important to understand Zoo Noise? As the word "Noise" indicates, some zoos and visitors block, distort, change, ignore, or interfere with the meaning of the message.

Students' experiences with nature are more likely to happen within their local surroundings during interactions with family and while viewing images. Therefore, zoos are a viable educational resource, and, more than ever, understanding how students' interactions during a zoo visit inform student learning is important. Zoos provide people with an opportunity to interact with new organisms, family, friends, strangers, and zoo personnel. This book identifies the interactions people have during a zoo visit and the importance of the conversations that take place during the visit. Because the Visitor Voice and Zoo Voice are equally important aspects of a zoo visit and their influences must be understood, research should be completed that evaluates the overlap of these voices. For example, what role do websites and new technology play in the Zoo Voice and how does this new technology influence the Visitor Voice?

By identifying the historical development of zoos from menageries to conservation education institutions, we have identified the evolution of exhibits from cages (for the curious to view exotic animals) to interactive, immersion exhibits. The exhibits of today incorporate cultural aspects and conservation ethics to inform responsible conservation citizenship. We cannot discuss zoos and their current roles and missions if we do not consider the historical past of zoos. As zoos evolved from

P.G. Patrick and S.D. Tunnicliffe, *Zoo Talk*, DOI 10.1007/978-94-007-4863-7_12, © Springer Science+Business Media Dordrecht 2013

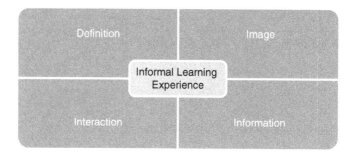

Fig. 12.1 Informal learning model

collections to conservation centers, they became modern indicators of conservation biology. The future of zoos and biological conservation may be bleak if the exhibits do not attract visitors. If people are not aware of why zoos exist, how they are useful, and how they aid in conservation, then the paying public may not be as supportive in the future. As we have situated our book in the historical literature and current research literature and proposed future research ideas, our book becomes current zoo history and shapes the future of zoos.

The Zoo Voice is one of biological information, and as many mission statements assert, it is a place for the conservation of flora/fauna. However, the Visitor Voice affirms that even though a zoo visit educates children, the visit is really to see animals. Visitors innately are interested in the animals without the additional interpretation provided by institutions. However, the specimen on display and the information included in the exhibit become the Zoo Voice and tell part of the conservation story. The Zoo Voice presents the organism as evidence that the species exists and as an exemplar of the species characteristics. Even though a popular belief is that industrialization and urbanization are reducing students' direct interactions with nonurban nature, we have discussed that alternative research does show that students are interacting with nature in their local environments. This is an important aspect when determining the Visitor Voice. Visitors do not visit the zoo *tabula rasa*. However, as the world becomes more urbanized, our personal experiences with animals become more isolated, in many cases limited to domesticated pets and urban species. No matter how zoos choose to get their voice heard by visitors, education is their most important biological and conservation function.

Because the Zoo Voice is a critical element in the contribution of zoos to conservation education, zoos need a comprehensive understanding of the knowledge visitors bring to the zoo—the Visitor Voice. By providing zoos the informal learning model (Fig. 12.1), we provide them with the theoretical underpinning for understanding the Visitor Voice and how visitors learn in an informal learning environment. Zoos serve an important role in allowing visitors to explore and join cultural identities by scaffolding concepts within their interpretation. Our work provides zoos and visitors with the theoretical underpinning for understanding the Visitor Voice and how visitors construct further understandings of animals in an

Fig. 12.2 Zoo knowledge model

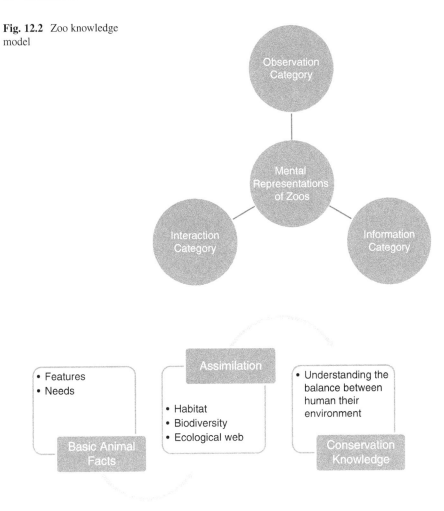

Fig. 12.3 Progression in acquiring conservation knowledge

informal learning environment. Furthermore, the zoo knowledge model (Fig. 12.2) defines the mental models people have regarding zoos and is an important tool that may be utilized when developing educational interactions. Zoo Noise is caused when the Zoo Voice fails to align with the zoo knowledge model and does not take into account the fundamental characteristics of the informal learning model. If zoos take into account the models described in this book, their education programs could be more powerful. The optimum way to address the conservation impacts of zoos and visitor education is to evaluate the divergent and nonscientific ways children construct their understandings of animals, ecology, and ultimately conservation. Biological conservation knowledge is constructed over time and is not necessarily linear (Fig. 12.3). However, the basic knowledge of what an animal is, including its features and needs, must be established before children are able to assimilate the ideas related to wildlife conservation such as habitat needs, biodiversity, and the organism's place in the ecological web.

Exhibits are an integral part of the Zoo Voice. Exhibits are designed by the zoo to encourage visitors to look at the organisms being displayed and to note important biological features. Zoos hope that by displaying organisms and providing information, the Zoo Voice will be heard by the visitor. However, between the visitor and the exhibit is an experiential space, in which the visitor may develop situational interest and construct further understandings. This experiential space is where the Visitor Voice and the Zoo Voice overlap instead of merging, which causes Zoo Noise. Within the experiential space, the zoo's exhibitry, interpretation, facilities, staff, and organisms and the visitor's prior knowledge and interest should join to scaffold the visitor's knowledge. However, the zoo exhibit/interpretation, a peer, or an adult may affect the scaffolding of knowledge. If the Zoo Voice and the Visitor Voice make a connection, the interactions within the zoo exhibitry may lead to situational interest, which leads to individual interest. During a zoo visit, the Visitor Voice consists of an inherent curiosity about animals and three personally constructed components of prior knowledge and interest: (1) understanding of the term "animal," (2) everyday taxonomy or categorization of animals, and (3) knowledge of the structures and behaviors of animals.

Zoos need to take into consideration that visitors come with their own agendas and knowledge. Visitors' agendas may not match the zoo's biological and conservation-oriented agenda. Therefore, the exhibit design should enhance the prior knowledge of visitors. In this book, we have attempted to identify and understand the conversations that take place in the exhibits' experiential space, explain why these interactions are important, and describe how zoo educators and teachers may take advantage of these learning opportunities. We have described research projects that have taken place in England and the USA, in which data was collected from students, ages 3–18, and adults with children. These studies were used to define the Visitor Voice by discovering what children look at in an exhibit, how adults respond and/or what it is they point out, and how students define zoos and the role of zoos. The main phenomena mentioned are the features of animals such as color, scent, shape, and size. Additionally, people are aware of babies, unusual characteristics, animal behaviors, movement, noises, location, and if it is feeding or moving. Determining the general topics people will discuss during a visit and their knowledge of zoos is an essential part in planning the educational components of exhibitry.

Why do people visit zoos in this technological age of multimedia and vicarious models of the living world? Because zoos are one of the environments in which people interact (though limited) with live animals. Schools visit zoos for curriculum learning, particularly biology and biodiversity, and families and other groups visit to see animals. Schools visit the zoo with predetermined agendas which usually are driven by worksheets or scavenger hunts. These predetermined agendas interfere with the students' ability to hear the Zoo Voice. Students are so busy running from exhibit to exhibit; they do not have an opportunity to spend time interfacing with the exhibitry. Moreover, this book recognizes the social influences and the need for partnerships between zoos, teachers, and universities. Planned, preparatory pre-visit, during visit, and post-visit activities will supplement or reinforce the visit. When zoo educators and classroom teachers work together to design educational materials, cross-curricular objectives including taxonomy, language arts, architectural studies, history, graphic representations (drawings), photography, art, and drama may be developed. Other visitors attend with a pre-visit agenda such as a social outing, stay-cation, date, day out, or birthday party. No matter what the reason for the visit is, educational or social, a visit should build on the visitor's knowledge of biology and allow the visitor to develop their concepts of taxonomy, behavior, and conservation.

The Visitor Voice begins before the zoo visit. Families discuss what they would like to see prior to the visit. Educational visits are driven by pre-visit decisions, which determine what children will see. In both cases, the knowledge the visitor brings to the zoo allows the visitor to interpret and explain the aspects of the exhibit (Zoo Voice). Effective educational visits are preplanned, become part of the lesson, and extend into the classroom post-visit. When teachers begin to plan for a zoo visit, they must include colleagues, zoo educators, and chaperones. Making sure the field trip stakeholders are aware of what should be learned and how these experiences will be elicited is important in making a field trip a successful learning experience. Chaperones must know the objectives, tasks, and timetable for the visit. Chaperones may not understand the learning cycle, but classroom and zoo educators must develop activities that include aspects of the learning cycle. Activities must be age appropriate and utilize the stages of learning exemplified by the learning cycle. Additionally, for classroom and zoo educators, assessing learning before, during, and after the visit is an important diagnostic tool that shapes how the activities are adapted. If student knowledge is not changing or improving, the activities must be adjusted. Utilizing assessment will provide proof for the zoo that a zoo visit is a worthwhile, educational event. Capturing and interpreting the changes in students' knowledge ensures that they are developing inquiry skills.

In closing, understanding where zoos have been and where they are going is an important part of ensuring the future of zoos. Research has been carried out concerning school visits to zoos, in-service teachers' zoo field trips, and preservice teachers' ideas concerning the educational benefits of zoo field trips. This book provides only a snippet of the complicated web of learning in zoos. Much is to be discovered about how people learn in zoos, how people view zoos, and how zoos may get their voice heard by the visiting public.

Index

P.G. Patrick and S.D. Tunnicliffe, *Zoo Talk*, DOI 10.1007/978-94-007-4863-7,
© Springer Science+Business Media Dordrecht 2013

CPSIA information can be obtained
at www.ICGtesting.com
Printed in the USA
LVHW080856201019
634733LV00014B/646/P